PENGUIN

HIPPOCRATIC WRITINGS

PROFESSOR SIR GEOFFREY ERNEST RICHARD LLOYD is Emeritus Professor of Ancient Philosophy and Science in the University of Cambridge and until 2000 he was Master of Darwin College. He was born in 1933 and educated at Charterhouse and Cambridge. His publications include *Polarity and Analogy* (1966), *Aristotle: The Growth and Structure of His Thought* (1968), *Early Greek Science* (1970), *Greek Science After Aristotle* (1973), *Magic, Reason and Experience* (1979), *Science, Folklore and Ideology* (1983), *Science and Morality in Greco-Roman Antiquity* (1985), *The Revolutions of Wisdom* (1987), *Demystifying Mentalities* (1990), *Methods and Problems in Greek Science* (1991), *Adversaries and Authorities* (1996), *Aristotelian Explorations* (1996) and *Greek Thought* (edited with Jacques Brunschwig, 2000).

HIPPOCRATIC WRITINGS

•

Edited with an introduction by
G. E. R. LLOYD

•

Translated by
J. CHADWICK AND W. N. MANN
I. M. LONIE
E. T. WITHINGTON

PENGUIN BOOKS

PENGUIN BOOKS

Published by the Penguin Group
Penguin Books Ltd, 80 Strand, London WC2R 0RL, England
Penguin Putnam Inc., 375 Hudson Street, New York, New York 10014, USA
Penguin Books Australia Ltd, 250 Camberwell Road, Camberwell, Victoria 3124, Australia
Penguin Books Canada Ltd, 10 Alcorn Avenue, Toronto, Ontario, Canada M4V 3B2
Penguin Books India (P) Ltd, 11 Community Centre, Panchsheel Park, New Delhi – 110 017, India
Penguin Books (NZ) Ltd, Cnr Rosedale and Airborne Roads, Albany, Auckland, New Zealand
Penguin Books (South Africa) (Pty) Ltd, 24 Sturdee Avenue, Rosebank 2196, South Africa

Penguin Books Ltd, Registered Offices: 80 Strand, London WC2R 0RL, England

www.penguin.com

First published by Blackwell 1950
This edition with additional material published
as a Pelican Classic 1978
Reprinted in Penguin Classics 1983

045

Introduction copyright © G. E. R. Lloyd, 1978
Translation of *Fractures* copyright 1928 by Loeb Classical Library
(Harvard University Press: Heinemann)
Translations of *The Seed, The Nature of the Child*
and *The Heart* copyright © I. M. Lonie, 1978
All other translations copyright 1950 by J. Chadwick and
W. N. Mann
All rights reserved

Printed and bound in Great Britain by Clays Ltd, Elcograf S.p.A.

ISBN-13: 978-0-14-044451-3

www.greenpenguin.co.uk

CONTENTS

CONTENTS

PREFACE

THE bulk of the Hippocratic Corpus is such that only a small proportion of the extant treatises can be included in a book such as this. The aim has been to present a selection of the most important and interesting treatises in the best available English translations. Many of the most famous treatises, including *The Oath, Tradition in Medicine* (otherwise known as *Ancient Medicine*), *The Sacred Disease,* and the first and third books of *Epidemics*, were translated by Dr J. Chadwick and Dr W. N. Mann in the volume entitled *The Medical Works of Hippocrates*, published by Blackwell, Oxford, in 1950. The translations in that volume, which are reproduced complete, with minor revisions, except for *Coan Prognosis*, form the major part of the present selection.

As the title of their book implies, the translations of Dr Chadwick and Dr Mann were of *medical* works. Translations of four of the outstanding non-medical treatises in the Hippocratic Collection have been added.

(1) The translation of *Fractures* by Dr. E. T. Withington, originally published in volume three of the Loeb edition of Hippocrates (Heinemann, London; Harvard University Press, Cambridge, Mass.) in 1928; *Fractures* is one of the two major surgical treatises (the other being *Joints*) in the Hippocratic Collection.

(2–3) Translations of *The Seed* and *The Nature of the Child* by Dr I. M. Lonie from the forthcoming Ars Medica series volume, containing edition, translation and commentary, by Dr Lonie and Professor G. Baader. Together with the work known as *Diseases* IV, *The Seed* and *The Nature of the Child* form the group of so-called embryological treatises that were almost certainly composed by the same author and that deal with the problems of generation, heredity and sex differentiation as well as with the growth of the embryo.

(4) A translation of *The Heart*: this, though later in date

7

than most of the other Hippocratic treatises, is the outstanding work dealing with an anatomical subject in the
Collection. In this case the translation by Dr I. M. Lonie has
been undertaken specially for this volume.

It is a pleasure to record my thanks first and foremost to
Dr Chadwick, Dr Mann and Dr Lonie, whose help and
counsel have greatly facilitated my work as editor; to Professor
Finley, who initiated the idea of such a volume; to my father,
Dr W. E. Lloyd, for advice on medical matters, and to my son,
Matthew, for help in preparing the index.

G. E. R. L.

Cambridge, February 1973

INTRODUCTION

In Western medicine, the name of Hippocrates has always stood for an ideal. Until comparatively recently in the history of Western medical thought, his views – that is, the views of the works that passed for his – were accepted as authoritative on all kinds of medical problems, and medical students read their Hippocrates not out of piety but as an essential part of their training as doctors. That is no longer the case: yet Hippocrates still represents an ethical ideal, the ideal of the compassionate, discreet and selfless doctor, and those who graduate from the medical schools of Scottish universities still, in 1973, subscribe to a version of his oath. Moreover, while the importance of Hippocrates has declined with the advance of modern medical knowledge, from another point of view the Hippocratic writings have not lost but gained in interest, as scholars have come to appreciate more fully their role in the development of Greek science, and in the subsequent history of Western science as a whole. Although Hippocrates is no longer assumed to be the repository of all medical wisdom, the importance of the writings associated with his name is threefold: first for the still living ethical ideal of the doctor that they represent, second for the insight they provide into the origins and development of rational medicine in the West, and third for the extraordinary influence that they exercised over medical thought over so many hundreds of years.

THE HIPPOCRATIC COLLECTION

The collection of medical writings known as the Hippocratic Corpus consists of about sixty treatises, some in several books, that vary widely in subject-matter, style and date. Although most of the treatises were written between 430 and 330 B.C., some are later works. The subjects covered include general pathology and the pathology of particular conditions,

9

diagnosis and prognosis, methods of treatment and of the preservation of health, physiology (the constitution of man), embryology, gynaecology, surgery and medical ethics. A few of the treatises are carefully composed lectures, some of which appear to be addressed to a general audience: that is, an audience that included laymen as well as medical students or doctors. Some other treatises too are fairly clearly defined entities, for example text-books dealing with a particular subject, or works arguing a particular thesis on a theme or themes announced at the outset. But a large number are composite productions, collections of diverse material, in some cases probably the work of several hands. Scrapbooks or notebooks such as the *Aphorisms* were not designed as unities and were often subject to additions and interpolations. Many of the Hippocratic treatises are practical manuals, and those who used them in the fifth and fourth centuries B.C. were, we may assume, far less concerned with such questions as the exact reading of the original text or the identity of the author or authors than with the substance of their contents, that is, the useful medical knowledge they contained.

The Corpus is evidently the work of a large number of medical writers, belonging to different groups or schools and representing in many cases quite opposed viewpoints, not only on such questions as the aetiology of diseases and the correct methods of treatment, but also on the methods and aims of medicine as a whole. Thus apart from the school of Cos, associated with Hippocrates himself, the rival school of Cnidus was probably responsible for several of the works in the Corpus, although widely differing views have been held on the identity of these Cnidian treatises.

Although numerous attempts have been made to identify, within the Corpus, the genuine works of Hippocrates himself – a problem that already exercised the ancient commentators on these writings – none can be said to have succeeded in this aim. The evidence available to us is poor and at points conflicting. Most of our detailed information about the life of Hippocrates comes from late and generally untrustworthy sources. Our earliest and most reliable authorities are Plato,

Aristotle, and Aristotle's pupil, the historian of medicine Meno. They tell us that Hippocrates came from Cos and that he was an approximate contemporary of Socrates (later sources date his birth more precisely, usually to 460 B.C.). They also prove that Hippocrates soon became famous as a doctor and they establish the not unimportant fact that he taught medicine for a fee, but they do not provide definite enough information concerning either his methods or his doctrines to enable us confidently to ascribe to him any one of the treatises in the Collection. In the absence of convincing arguments for authenticity, those who have discussed the problem have all too often fallen back on the supposition that Hippocrates must be the author of the treatises that they happen to value most highly. The Hippocratic treatises, as they have come down to us, are all anonymous. It is possible that they contain some of Hippocrates' own work. But there is no means of establishing with certainty, or even with a high degree of probability, that this is so or, if so, which work is his. The one treatise whose authorship can be settled with some confidence is *The Nature of Man* (or at least the major part of that composite work), and the author in question is not Hippocrates but his son-in-law Polybus. It must, however, be stressed that the importance of the Hippocratic Collection is independent of any connection with Hippocrates the man.

While individual treatises are occasionally referred to in closely contemporary writings, the Collection as such cannot be traced back before the work of the first commentators and lexicographers in the third century B.C. The suggestion that it was put together, in the main, by scholars working at Alexandria at that time offers perhaps the most likely explanation of its origin. But once the nucleus of the Collection existed and had become associated with the name of the most famous doctor of the classical period, it was no doubt subject to later additions and exercised a powerful attraction on other anonymous medical literature, including some works that were evidently composed much later than the bulk of the Corpus. Thus of the treatises included in this selection, *The Heart* has been dated, on the grounds both of style and of content, after Erasistratus of Ceos, an important Hellenistic doctor and biologist who was active about 260 B.C. A study of the ancient commentators shows that there were always considerable variations in what passed for Hippocrates, and this is also reflected in the fact that the treatises contained in our different manuscript traditions for the Hippocratic texts vary appreciably. Despite the scholarly attention that these works received in antiquity, no single authorized canon of Hippocrates was established.

Although the variety of views represented in the Hippocratic writings, and the complexity of the interrelations of different treatises or groups of treatises, are a source of some dismay to those who seek to identify the work of a particular author or authors, the controversies reflected in the Corpus provide precious information concerning the growth of rational medicine in Greece. Evidently many fundamental early Greek medical ideas were formed in the crucible of the type of debate exemplified in the Corpus – debates in which not only particular theories and practices but also the whole question of the nature and aims of medicine itself were discussed. Indeed the disputes on such topics, both within the medical profession and between the doctors on the one hand and the philosophers on the other, illuminate not only the development of medicine but also that of science as a whole in Greece. Quite apart from the wealth of information about early Greek medicine that we gain from the Hippocratic writings, they provide, as we shall see, invaluable first-hand evidence relating to certain crucial aspects of the growth of scientific inquiry in early Greek thought.

THE MEDICAL PROFESSION IN THE FIFTH AND FOURTH CENTURIES

The Hippocratic writings stand at the beginning of systematic medical inquiry in Greece. To be sure, the Greeks themselves liked to trace the origins of medicine back through the heroic figures of Machaon and Podalirius, mentioned in Homer, to the mythological divine founder Asclepius, and we have no reason to doubt that many of the treatments employed from the sixth century onwards, including such a difficult and dangerous one as the practice of trephining, had been used not only in Greece itself, but also in other areas of the Near East,* long before. Yet it was in the late sixth or early fifth century that the first sustained critical investigations into the causes and treatment of diseases began and that we find the first attempts to define and defend the status of medicine as a rational discipline or *technē*.

It is important to recognize at the outset, however, just how precarious the practice of medicine was at the time when the Hippocratic authors were writing. Although we may speak loosely of those who engaged in medical practice full-time as professional doctors, medicine was not a profession in the fullest modern sense of that term. The essential point is that, unlike his modern counterpart, the ancient doctor possessed no legally recognized professional qualifications. Anyone could claim to heal the sick, and the doctors were in competition not only with midwives, herbalists and drug-sellers, but also with the type of 'purifiers' and sellers of charms and incantations who are criticized and rebutted in *The Sacred Disease*. Again, the distinction between the doctor and the gymnastic trainer was sometimes a fine one, and experience in the gymnasia was an important part of the training of many of those who practised medicine and surgery in ancient Greece.

The insecurity felt by doctors about their own position is evident in the way the Hippocratic writers frequently find it necessary to uphold the claim that there is such a thing as

*The question of the debts of Greek medicine to the medicine of their ancient Near Eastern neighbours, particularly the Egyptians, is obscure and controversial. The famous Edwin Smith papyrus, which dates from around 1600 B.C. but incorporates much earlier material, shows that detailed case-histories had begun to be collected in Egypt long before medicine became established in Greece. It has also recently been argued that the Cnidian school of Greek doctors in particular derived certain pathological doctrines from Egypt. Yet though the remains of Egyptian, Assyrian and Hittite medicine are impressive, there is nothing comparable with the systematic debates on, for example, the causes of diseases and the nature of medicine itself that we find in the Hippocratic Corpus. They, so far as we can judge, were a new phenomenon.

medicine and that it is a genuine *technē*, the Greek term whose range covers both what we should call arts and crafts and what we should call sciences. Indeed the whole of the treatise *The Science of Medicine* is devoted to this theme and to defending medicine against its detractors. In chapter 3, for instance, the writer says: 'First of all I would define medicine as the complete removal of the distress of the sick ... It is my intention to prove that medicine does accomplish these things and is ever capable of doing them. And as I describe the science I shall at the same time disprove the arguments of her traducers.' Other Hippocratic authors too repeatedly insist on the distinction between the doctor and the mere layman* and on that between the doctor and the quack or charlatan. Yet the doctor had no formal qualification to point to in support of his claim.

As in the case of other arts, those who wished to practise medicine normally received their elementary training by becoming attached as assistants to established doctors. It helped to have been associated with the doctors at one or other of the best known centres of medical training. Already in the late sixth century, two Greek city-states, Croton (in Magna Graecia, now southern Italy) and Cyrene (in North Africa), were famous for their doctors, and in the fifth century, Cos, the birthplace of Hippocrates, and Cnidus, just opposite it on the mainland of Asia Minor, developed flourishing medical schools, in both senses of the word 'school': they became the main centres for the teaching of medicine, and the doctors associated with either place shared certain medical doctrines and practices.† Yet while the concentration of medical men

---

* *idiōtēs*, the word from which the English 'idiot' is derived.

† Some characteristic Cnidian doctrines can be identified, thanks to the criticisms of them that we find in *Regimen in Acute Diseases* and in Galen's commentary on that work. Thus according to Galen the Cnidians went in for fine distinctions in their classification of diseases, identifying no less than twelve separate diseases of the bladder and four of the kidney; and it appears that bile and phlegm played a particularly prominent role in their aetiology of diseases. Their therapeutics are explicitly criticized in *Regimen in Acute Diseases* on the grounds that they used too few remedies: 'They generally prescribe opening medicine and recommend their patients whey and milk to drink.'

in such places was obviously advantageous to the student, instruction was still, one must assume, largely informal. There was certainly no set term to the period of training that a medical student should undergo, nor at the end of it did he obtain certification of his right to practise. His establishing himself as a doctor depended less on how he had been trained than on the reputation he acquired in practice, and where the whole question of his basic competence might be disputed at any stage, the problems he faced in winning and keeping the confidence of his clients were formidable.

In general Greek doctors practised privately, but occasionally we hear of doctors being employed, usually for a year at a time, by a city-state. Evidence for the institution of public doctors goes back to the sixth century, but their role and duties are disputed, and it is not clear precisely how extensive this institution was. It is unlikely that the intention of such appointments was to provide a free state health service for all, or even for all the citizens, and it may be that all that the state required and paid for was that the doctor should reside and practise in the city. To be sure, the late Hippocratic treatise *Precepts* (ch. 6) recommends that doctors should, where necessary, be prepared to treat their patients without payment, and we may assume that this sometimes happened. But the usual rule was that doctors charged their patients for their services, no doubt adjusting their fees to their patients' means, and it is possible that this is what the public doctors also did.

While some doctors were permanently resident in a particular city, a large number travelled from place to place in search of a living and in response to the varying demand for their services. One of the treatises in our selection, entitled *Airs, Waters, Places*, is a manual whose chief purpose is to help the itinerant doctor to anticipate the different types of diseases that are likely to occur in cities with different geographical and physical conditions. The writer distinguishes, for example, the diseases that are likely to be common in cities exposed to the north winds, in those facing south, in those where the water supply comes from stagnant sources such as marshes and in those where it comes from rock springs. The writer's presuppositions are highly schematic, but the problem of adapting to local conditions was a real one for the itinerant practitioner.

The comparatively insecure position of the doctor is reflected in many features of Greek medical practice. Thus one of the explicit aims of the practice of diagnosis and prognosis is to impress the patient and win his confidence. The work entitled *Prognosis* indicates that the Greek doctors tried to tell their patients not only what was going to happen to them but also their present and past symptoms. The author recommends (ch. 1) that the doctor should fill in the details that the patients themselves have omitted: he will thereby 'increase his reputation as a medical practitioner and people will have no qualms in putting themselves under his care'. The practice of prognosis was evidently an important psychological weapon in the battle to win the patient's confidence.

Similar preoccupations underlie the advice, given in several Hippocratic texts, not to undertake hopeless cases. It is true that another point of view is also sometimes expressed, namely that however hopeless the case the doctor should do whatever he can to help the patient; and the authors of the works known as the *Epidemics*, for example, clearly had no compunction in admitting that a large number of the cases they described – and that were presumably under their care – ended in death. Indeed in books I and III, the majority of the case-histories described (25 out of 42) have a fatal outcome. Yet apart from considerations of honesty, the need to protect one's reputation was with some writers an important factor. Thus *The Science of Medicine* (ch. 3) even takes as a defining characteristic of medicine 'the refusal to undertake to cure cases in which the disease has already won the mastery, knowing that everything is not possible to medicine'. The author of *Prognosis* (ch. 1) is also aware of the problem: 'By realizing and announcing beforehand which patients were going to die, he would absolve himself from any blame.'

Even more striking are the views expressed by the author of *Fractures* on the dilemma the doctor faces. Describing the reduction of the thigh and the upper arm, the writer notes (ch. 35) that 'in such injuries ... one must not overlook the dangers or the nature of some of them, but foretell them as suits the occasion'. But then in the next chapter, although he gives more information concerning the treatment of such cases – on, for example, the diet to be prescribed – he ends: 'One should especially avoid such cases if one has a respectable excuse, for the favourable chances are few, and the risks many. Besides, if a man does not reduce the fracture, he will be thought unskilful, while if he does reduce it he will bring the patient nearer to death than to recovery.' The selflessness and dedication shown by many Greek doctors can be seen not only in such works as the *Epidemics*, but also in, for example, Thucydides' account of the plague at Athens (II, 47ff.) - where he notes the high incidence of mortality from the disease among the doctors who attempted to treat it. Yet the relationship of trust between patient and doctor was a fragile one, and this is reflected in the doctors' anxieties concerning the repercussions of any apparent failure.

On the question of the civil status of those who practised medicine, a passage in Plato's *Laws* has been taken to suggest that there was a clear distinction between the clientèles of doctors who were themselves slaves and of those who were free men. At *Laws* 720 cd he speaks of free doctors treating the free and of slave doctors treating their fellow-slaves. Yet this evidence must be treated with caution. We must certainly take it that there were slaves among those who were called doctors and who practised medicine in the ancient world. It seems unlikely that any of the Hippocratic authors was not free, but we have simply no direct information on this point. Yet in one respect, at least, the evidence of Plato is contradicted by what we find in the Hippocratic Corpus. The patients whose case-histories are recorded in the *Epidemics* include representatives from all walks of life, rich and poor, citizens, slaves and visitors from abroad.

The social status of doctors no doubt also varied a good deal. Despite the fact that Hippocrates taught medicine for money, there is a striking contrast between the way he is referred to in Plato and the way Plato treats most of the professional educators or sophists. In Plato's *Protagoras* (311b ff.) Hippocrates is on a par with the great sculptors Polycleitus and Pheidias. In the *Phaedrus* (270c ff.) Hippocrates is first mentioned by Phaedrus, but then Socrates proceeds to equate what Hippocrates has to say with the true account of the matter, whether this is out of deference to Hippocrates, or to Phaedrus, or to both of them. The way in which the doctor Eryximachus is portrayed in Plato's *Symposium* is also revealing. Eryximachus is clearly no armchair medical theorist, but a man of experience, who is keen to show his respect for his art. Though at times rather pompous, he is depicted on the whole sympathetically, and the seriousness of his treatment of the theme of love helps to pave the way for the climax of the dialogue in Socrates' speech. Anyone who earned money by practising a skill was liable to be treated as a social inferior by men whose leisure was guaranteed by inherited, usually landed, wealth, and this was often the fate of doctors in Greece, and more especially later in Rome. Yet at Athens in the classical period Plato represents Eryximachus (and one must presume there were many doctors like him) as associating on equal terms with the other guests at the symposium, who include the poet Aristophanes and the statesman Alcibiades.

It is clear that, despite the hazards of medical practice, some doctors were highly successful and earned large sums of money. This can be seen, for example, from the story of Democedes of Croton, reported in Herodotus (III, 129ff., especially 131). Democedes was employed as a public doctor in three successive years, by the city-states of Aegina and Athens and by Polycrates of Samos, and each year his salary increased, being first one talent, then 100 minae, then two talents. The value of these sums can be judged from the fact that the normal daily wage of a skilled worker in the late sixth and early fifth centuries was a drachma – there being 100 drachmae to the mina and sixty minae to the talent. Nor is it likely that Democedes' salary as a public doctor was his only source of income. He may well have had fees from some of his patients in addition, and, as we have seen, it was also possible for a doctor to earn money by teaching medicine, whether to students intending to practise medicine themselves or (as I shall indicate later) to people who wanted to learn something about medicine simply as part of their general education. But Democedes' case was no doubt an exceptional one: Herodotus notes that the fame of the doctors of Croton was largely due to him. The income of ordinary doctors must have ranged within wide limits according to their skill and reputation and according to the varying demand for their services.

Several of the Hippocratic treatises that deal with questions of medical etiquette and ethics warn the doctor against avarice. *Precepts* (ch. 6) recommends that the doctor should consider the patient's means in fixing fees and, as already noted, suggests that the doctor should be prepared, on occasion, to treat a patient for nothing. The same work also says (ch. 4) that the doctor should not begin a consultation by discussing fees with his patient. This may well cause the patient anxiety, for he may believe that the doctor will abandon him if no agreement over fees is reached. As the writer puts it: 'It is better to reproach patients you have saved than to extort money from those in danger of dying.' *Decorum* (ch. 5), too, mentions lack of the love of money as one of the qualities a good doctor should show. The treatises in which these sentiments are expressed are later than the bulk of the Hippocratic Corpus and belong to the period after Aristotle. Later still, we find plenty of complaints, in both Greek and Roman writers, concerning the greed and cupidity of doctors who are sometimes described as making colossal fortunes from their gullible patients. One such writer is Pliny (first century A.D.), who goes into the topic at length in book 29 of his *Natural History*. No doubt he and others who dwelt on this theme are guilty of some exaggeration, but it is evident that from the end of the fourth century B.C., at least, it was quite commonly believed both that some doctors were avaricious and that large sums of money were to be earned by the most successful practitioners.

Although the boundary between doctor and layman was far

from being clearly defined, this did not prevent attempts being made, in some quarters at least, to limit and control those who became doctors. One text that is relevant to this point is the Hippocratic *Oath*. The questions of who took this oath in the ancient world and how representative were the ideas expressed in this work are hotly disputed. Certainly some of the specific injunctions, for example not to operate 'even for the stone', run counter to common Greek medical pratices of the fifth and fourth centuries B.C., and it has therefore been thought likely that, although many of its ideals and beliefs were widely shared, *The Oath* as such belongs to a particular group of practitioners rather than to Greek doctors as a whole. But on the specific question of who should be taught medicine, *The Oath* is explicit: 'I will hand on precepts, lectures and all other learning to my sons, to those of my master and to those pupils duly apprenticed and sworn, and to none other.' Evidently this group of doctors, at least, set up rules both to govern who was to be allowed to receive medical instruction and to make sure that these rules themselves were handed on to each successive generation of students.

Finally, one further manifestation of some of the pressures on the medical profession is the phenomenon of intentional obscurity in some medical writings. In many cases, to be sure, our difficulty in understanding Greek medical literature stems from other causes, such as the corruption of the text or the lack of background knowledge. But it seems clear that in some works obscurity of expression has been deliberately cultivated. This is particularly true of such treatises as *Nutriment* and *Humours*, whose style is not just pithy, but opaque, oblique and elliptical, and which appear to have been composed with a particular esoteric audience in mind. In the ancient world not only the initiates of the mystery religions, but also, for example, the early followers of Pythagoras, were required, in their different contexts, to practise secrecy. To judge from the frank character of most of the extensive extant medical literature of the fifth and fourth centuries, Greek doctors were in general quite open about their ideas, practices and discoveries. Yet we have seen that medicine was both a competitive and a precarious calling, and some doctors not only formed closely-knit sects, but were also positively secretive about their art, and prepared to write about it only in veiled language that none but the initiated would understand. Thus as *Decorum* (ch. 18) puts it, cryptically enough, 'things that are glorious are closely guarded among all men', while the *Canon* ends in terms that deliberately echo those of the mystery religions: 'Holy things are revealed only to holy men. Such things must not be made known to the profane until they are initiated into the mysteries of science.'

HIPPOCRATIC MEDICAL THEORIES AND PRACTICES

Greek ideas on the causes of diseases were extremely varied and it should be emphasized at the outset that there is no such thing as the one, nor even the single dominant, Hippocratic medical doctrine. Theories on the subject ranged all the way from the belief that all diseases have a single origin to the view that there are as many different diseases as there are patients, or that wherever any difference whatsoever can be found between two sets of symptoms, two different diseases must be diagnosed. Controversy also raged, both between different schools and between individuals of the same school, on the question of the nature of the causal factors at work. All we can hope to do here is to outline some of the more common and important notions.

Despite considerable differences on the explanations of diseases, there was a wide measure of agreement among the Hippocratic doctors on the names and descriptions of the main kinds of condition. In general, conditions were identified either by the part of the body affected, or by the most prominent signs or symptoms. Examples of the first are *nephritis* (from *nephros*, kidney), *hēpatitis* (from *hēpar*, liver), *pleuritis* (cf. our pleurisy, from *pleura*, rib or side), *arthritis* (from *arthron*, joint) and *ophthalmia* (from *ophthalmos*, eye), examples that also illustrate the longevity of Greek medical terminology, although nowadays the same, or derived, term may not be used to refer to the same condition, or may be used to refer to a similar condition now understood quite differently. Examples of the second kind are *phthisis* (consumption, literally wasting), *hydrōps* (dropsy, from *hydōr*, water), *tetanos* (from *teinō*, stretch) and *strangouria* (strangury, from *stranx*, trickle, drop, and *ouron*, urine).

Many of the most common Greek terms cut across modern medical categories and are strictly untranslatable. One instance noted by Drs Chadwick and Mann is the word *causus*. As they observe, in some, but not all, cases enteric fever is clearly described, but the term is used of a variety of fevers which we now separate into a number of different diseases. Fevers in general were classified by the Greeks according to their observed or imagined periodicities. Thus tertians were fevers where a 'crisis' or marked change in the symptoms occurred every third day by Greek reckoning (that is, every other day by ours: the Greeks counted both the first and the last day). Others were termed quartans, quintans and so on, and those that did not manifest obvious or definite periods were called 'disorderly' or 'wandering'. Where, as often, the Greeks identified diseases from signs and symptoms, there is no hard and fast distinction between the term used descriptively of the sign itself, and the same term used inferentially, implying an interpretation or diagnosis of the disease in question.

Disease was generally seen as some sort of imbalance in, or disturbance of, the natural state of the body, and the notion of diseases being hostile to nature or to the body runs through many Greek medical writers of different theoretical persuasions. The doctor's role was to combat the disease or to help nature to do so. This idea of a war between the disease on the one hand and the doctor and nature on the other was, however, associated with many different explanations of the origins of diseases.

One way of subdividing the causal factors in diseases was into internal and external ones. *The Nature of Man* (ch. 9) has this argument: 'When a large number of people all catch the same disease at the same time, the cause must be ascribed to something common to all and which they all use; in other words to what they all breathe ... However, when many different diseases appear at the same time, it is plain that the regimen [that is, diet and exercise] is responsible in individual cases.' Different writers refer to a wide variety of external and internal factors, whether or not they accept that argument. Thus among external factors, apart from the air we breathe, the winds and the seasons generally were thought to cause diseases. The north and south winds in particular are often represented as being especially influential. The author of *Aphorisms* writes (Sec. III, ch. 5): 'South winds cause deafness, misty vision, headache sluggishness and a relaxed condition of the body ... The north wind brings coughs, sore throats, constipation, retention of urine accompanied by rigors, pains in the sides and breast.' And the author of *The Sacred Disease* believes that epilepsy is more likely to occur when there is a change in the wind and he devotes chapter 16 to describing the effects of the north and south winds both outside and inside the body.

Particular diseases were often correlated with particular seasons or weather conditions. In the 'constitutions' preceding the case-histories, the books of *Epidemics* contain detailed descriptions of the weather conditions accompanying outbreaks of diseases, and many writers suggest that abnormal weather or sudden changes in it cause diseases. *Airs, Waters, Places*, which aims to provide information about the seasons that will enable the doctor to foretell how the year will turn out, emphasizes that the successful practice of medicine depends on knowing 'what changes to expect in the weather' (ch. 2), and the author remarks (ch. 11) that 'it is particularly necessary to take precautions against great changes'. Again, *Aphorisms* states the common doctrine generally (Sec. III, ch. 1): 'The changes of the seasons are especially liable to beget diseases, as are great changes from heat to cold, or cold to heat in any season. Other changes in the weather have similarly severe effects.'

The internal factors most commonly mentioned are diet and exercise (both of which are included in the term 'regimen'). Several writers develop a theory of the need to balance these two. Thus the author of *Regimen* I (ch. 2) states that food and exercise have opposite powers but that they both work together to produce health, and it was often held that diseases are caused by excess either in overeating or in fasting – or more generally by any state of 'repletion' or fullness, or 'depletion' or emptiness.

Regimen I and other treatises maintain that eating the wrong foods, or too much or too little food, disturbs the balance of the natural constituents of the body, but there was as little agreement among the medical writers as to what these constituents are, as there was among the philosophers concerning the more general question of the fundamental elements of all physical bodies. *Regimen* I claims that the constituents of all things, including man, are fire (thought of as hot and dry) and water (thought of as cold and wet). In this the author is obviously influenced by, and remains close to, pre-Socratic physical speculation. The philosopher Empedocles, in particular, was responsible for the first clear statement of what was to become one of the most important and influential physical theories in antiquity, namely that all things consist of earth, water, air and fire, although this theory had many rivals, especially in the various ancient versions of atomism.

As is clear not only from the theories that are either stated or attacked in the Hippocratic Corpus, but also from the evidence of the papyrus Anonymus Londinensis, which preserves fragments of the history of medicine written by Aristotle's pupil Meno, a great variety of physical theories based on one or more of the four Empedoclean elements, or on one or more of the four primary opposites, hot, cold, dry and wet, were current among medical writers in the late fifth and early fourth centuries. But some doctors developed physical theories concerning the elements in the body, and corresponding pathological doctrines concerning the origins of diseases, in terms of other factors, for example 'the sweet', 'the bitter', 'the acidic', 'the astringent', 'the insipid' and so on; these too, like 'the hot', 'the cold', 'the dry' and 'the wet', were generally conceived not so much as what we should call qualities, but as things – the sweet (or hot) substance or principle or element. Connected with these, in turn, there

were yet other doctrines based on one or more of the commonly recognized humours (*chymoi*: the primary meaning of the
term is juice or flavour). The two most important of these
were bile and phlegm, which play a particularly prominent
part in pathological theories. Yet here again there was no
agreement among the medical writers either about how many
the principal humours were, or about their origin and role, for
some writers maintained that they are natural or congenital,
others that they are pathological, and some saw them as the
causes, others as the products, of diseases.

Two examples will illustrate one type of pathological theory
common in the Hippocratic Corpus. The author of the treatise
called *Affections* (ch. 1) asserts: 'In men, all diseases are caused
by bile and phlegm. Bile and phlegm give rise to diseases when
they become too dry or too wet or too hot or too cold in the
body'; and he goes on to say that such changes are brought
about in a variety of ways, for example by food and drink,
exercise, wounds, 'smell, hearing and sight', sexual intercourse, and 'the hot' and 'the cold' themselves. Again the
treatise *Diseases* I (ch. 2) states: 'All diseases come to be, as
regards things inside the body, from bile and phlegm, and as
regards external things, from exercise and wounds, from the
hot being too hot, the cold too cold, the dry too dry and the
wet too wet. And bile and phlegm are formed in things as they
come to be and they always exist, in greater or lesser quantities,
in the body, and they bring about diseases, both those arising
from food and drink and those from excess of hot and cold.'

As such examples indicate, Hippocratic statements concerning the origin of diseases often take the form of very
sweeping generalizations indeed. Several writers develop
complex doctrines covering not only the causes of diseases and
the constituents of the body but other matters such as the
cycle of the seasons, elaborating their theories far beyond
their immediate relevance to the problems of the diagnosis
and the treatment of disease. One of the most striking examples
is in the work *The Nature of Man*. This begins (chs. 1–3) by
refuting physical and physiological theories based on a single
element, whether air, fire, water or earth, or again blood,

bile or phlegm. If there were only one element, the writer argues, generation would not be possible, nor would we experience pain. His own theory of what the body is composed of draws heavily on earlier philosophical ideas. Yet the synthesis he proposes is, in certain respects, more elaborate and comprehensive than any extant earlier system, notably in the correlations he suggests between the four primary opposites, the four basic humours and the four seasons. The basic constituent substances, in terms of which all physical bodies are to be analysed, are the hot, the cold, the dry and the wet. Thus 'each of the elements must return to its original nature when the body dies; the wet to the wet, the dry to the dry, the hot to the hot and the cold to the cold' (ch. 3). But the human body may be thought of as composed, in the first instance, of the humours blood, phlegm, yellow bile and black bile. As he writes in chapter 4: 'These are the things that make up its constitution and cause its pains and health. Health is primarily that state in which these constituent substances are in the correct proportion to each other, both in strength and quantity, and are well mixed. Pain occurs when one of the substances presents either a deficiency or an excess, or is separated in the body and not mixed with the others.' Moreover each of the humours is associated with one of the four seasons (blood, yellow bile, black bile and phlegm are said to predominate in turn in the body in the four seasons, spring, summer, autumn and winter respectively, ch. 7) and with two of the primary opposites, hot, cold, dry and wet. Thus yellow bile, like summer, is dry and hot. The doctrine is nothing if not neat and systematic, and one of the attractions of such theories is that they are so simple and yet so all-embracing. Certain empirical evidence is adduced and incorporated into the theory: for example, the writer mentions that phlegm is found to be cold to the touch and he refers to the effects of drugs that reveal the presence of humours in the body. But the theory is developed as an abstract schema, as a highly speculative, and would-be comprehensive, physical, physiological and pathological doctrine.

But while there are marked speculative tendencies in several

Hippocratic writers, others are strongly critical of those same tendencies. The most remarkable document in this respect is the treatise *Tradition in Medicine*. This castigates theorists who use what the writer calls hypotheses or postulates in the explanation of diseases, and he is especially critical of theories based on hot, cold, dry and wet in particular. His first quarrel with his opponents is that they narrow down the causes of diseases (ch. 1): 'They have supposed that there are but one or two causes; heat or cold, moisture, dryness or anything else they may fancy.' Again, medicine is an art in which there are both good and bad practitioners. It is not like such subjects as astronomy and geology which deal with 'invisible or problematic substances' and where, as he puts it, 'a man might know the truth and lecture on it without either he or his audience being able to judge whether it were the truth or not, because there is no sure criterion.' As for the hot and the cold and so on, the writer has this to say (ch. 15): 'I am utterly at a loss to know how those who prefer these hypothetical arguments and reduce the science to a simple matter of "postulates" ever cure anyone on the basis of their assumptions. I do not think that they have ever discovered anything that is purely "hot" or "cold", "dry" or "wet", without it sharing some other qualities.' Indeed (ch. 16) the hot and the cold are 'the weakest of the forces which operate in the body'.

Yet critical as he is both of theories based on hot, cold, wet and dry, and of the use of arbitrary postulates in general, the writer himself does not, of course, manage to do without certain assumptions in his own theories. Indeed his particular suggestions concerning the constituents of the body bear, we should say, obvious similarities to those that he had singled out for special condemnation. Admittedly his own doctrine is rather more complex in that he postulates a large number of constituent substances in the body. But he describes those substances as follows (ch. 14): 'There exists in man saltness, bitterness, sweetness, sharpness, astringency, flabbiness and countless other qualities having every kind of influence, number and strength. When these are properly mixed and compounded with one another, they can neither be observed nor are they harmful. But when one is separated out and stands alone it becomes both apparent and harmful.' The writer wishes to exclude all 'postulates' from medicine, but himself employs assumptions similar in type to those he criticizes: indeed it is apparent that, at this period, *any* attempt at a general theory concerning the constituents of the body and the origins of disease was bound to be based on what might be described, in the writer's terms, as unverified hypotheses.

The Greek doctors had, and at this stage in the history of medicine could have, no knowledge of many of the causal factors relating to the conditions they encountered. In particular they had, of course, no conception of the role of microorganisms in causing diseases. The theories they produced provided a framework into which could be fitted many of their observations, for example of temperature changes, of periodicities, and of discharges from the body. But their general pathological doctrines were all more or less speculative, and more or less arbitrary, conjectures. Yet as we have seen, disagreements among Greek medical writers related not only to first-order questions, concerning what the causes of diseases are, but also to second-order ones, concerning both the type of theory and the nature of the support required for it, and this provides important evidence of a growing interest in questions not just of medical, but of general scientific, method.

Among the medical writers who adopted a cautious and sceptical view of generalizations about the causes of diseases is the author of *Tradition in Medicine*. Although he is prepared to advance general theories himself, he recognizes that the science of medicine is inexact (ch. 9): 'One aims at some criterion as to what constitutes a correct diet, but you will find neither number nor weight to determine what this is exactly, and no other criterion than bodily feeling. Thus exactness is difficult to achieve and small errors are bound to occur. I warmly commend the physician who makes small mistakes; infallibility is rarely to be seen.' Doubt about how much the doctor can know is expressed by other writers too. In practice, the

principal problem that the doctors faced – whether or not they
had a general pathological theory – was that of isolating the
relevant factors in the diseases with which they had to deal.
Even though there was as yet no technical vocabulary to draw
the distinction between causal and coincidental factors, the
medical writers are aware of the problem of distinguishing
what is, and what is not, responsible for any given condition.
Thus *Regimen* III (ch. 70) remarks that 'the sufferer always
lays blame on the thing he may happen to do at the time of the
illness, even though this is not responsible'. *Tradition in Medicine*
notes more generally: 'Most doctors, like laymen, tend to
ascribe some such event [a complication in a disease] to some
particular activity that has been indulged in. In the same way
they may ascribe something as being due to an alteration in
their habits of bathing or walking or a change of diet, whether
this is the actual case or not. As a result of jumping to conclusions, the truth may escape them' (ch. 21). Chapter 19 of
the same work reveals one of the criteria that the writer
assumes a cause must fulfil: 'The cause of these maladies is
found in the presence of certain substances, which, when
present, invariably produce such results.' Other treatises state
such points as that 'every phenomenon will be found to have
some cause' (*Science of Medicine*, ch. 6) or that 'each disease has
a natural cause and nothing happens without a natural cause'
(*Airs, Waters, Places*, ch. 22); some recognize that a similar
effect can be brought about by different causes (for example,
Regimen in Acute Diseases). Greek thought on the topic of
physical causation was, in general, slow to develop, only
gradually building up a vocabulary of terms that were usually
derived from the sphere of human responsibility and that in
many cases retained their social or even political associations.
Thus before *aitia* came to be used generally in the sense of
'cause', it meant responsibility or blame, and the meaning of
to aition is equivalent to 'that which is responsible'. When the
Hippocratic authors write of *dynameis* at work in the body, the
term sometimes retains some of the sense of 'political power'
alongside that of 'physical force' – as we can judge from its
use in conjunction with other cognate terms such as *dynasteuō*

(reign), in which the political connotation clearly predominates.* So far as the inquiry into nature is concerned, it so happens that both the distinction between cause and coincidence, and the idea that every effect has a cause, are first clearly expressed in medical writers, and although we should make due allowance for the fact that much more fifth century medicine than fifth century cosmology has been preserved, it is clear that the investigation of diseases provided one of the chief contexts in which Greek ideas on physical causation developed.

When first called in to a case, the doctor aimed, as we have seen before, to give a 'prognosis' covering the past, present and future of the disease. Although the doctor may well have named the condition he thought the patient was suffering from, he no doubt concentrated, in practice, less on giving the patient a theoretical explanation of the cause of the disease than on the question of its outcome and particularly on the chances of recovery. There can be no doubt, however, that the Greeks held that the prognosis should be based on a very thorough examination of the patient. In *Prognosis*, which is particularly concerned with 'acute' diseases, that is those accompanied by high fever such as pneumonia or malaria, the writer gives detailed instructions about how the doctor should proceed. First he should examine the patient's face, for example the colour and texture of the skin, and especially the eyes, where he should consider whether 'they avoid the glare of light, or weep involuntarily', whether 'the whites are livid', whether the eyes 'wander, or project, or are deeply sunken', and so on. He should also inquire how the patient has slept, about his bowels and his appetite: he should take into account the patient's posture, his breathing and the temperature of the head, hands and feet, and separate chapters are devoted to how to interpret the signs to be found in the patient's stool, urine,

*E.g. *Tradition in Medicine*, ch. 16. The relation between different factors in the body was often conceived with the help of political metaphors or images, especially that of the balance of power; this image goes back to Alcmaeon, who is reported to have held that health lies in the *isonomia* or equal rights of certain powers (*dynameis*) in the body.

vomit and sputum. The one notable absentee from the list of
things the doctor should consider is the pulse. Although the
phenomena of pulsation, throbbing and palpitation are referred to by Hippocratic writers, the value of the pulse in
diagnosis was not appreciated until after the date of most of
the Hippocratic treatises. The first person to restrict the pulse
to a distinct group of vessels, the arteries, and to recognize
that it can be used as an indicator of disease was Praxagoras of
Cos, working about 300 B.C.

Another detailed account of the factors that the doctor
should consider in diagnosing appears in *Epidemics* I (ch. 23),
a passage that is worth quoting at length. After saying that
one should take into account 'the nature of man in general
and of each individual and the characteristics of each disease',
the writer proceeds:

Then we must consider the patient, what food is given to him
and who gives it . . . , the conditions of climate and locality both in
general and in particular, the patient's customs, mode of life,
pursuits and age. Then we must consider his speech, his mannerisms, his silences, his thoughts, his habits of sleep or wakefulness
and his dreams, their nature and time. Next, we must note whether
he plucks his hair, scratches or weeps. We must observe his
paroxysms, his stools, urine, sputum and vomit. We look for any
change in the state of the malady, how often such changes occur
and their nature, and the particular changes which induce death or
a crisis. Observe, too, sweating, shivering, chill, cough, sneezing,
hiccough, the kind of breathing, belching, wind, whether silent or
noisy, haemorrhages and haemorrhoids. We must determine the
significance of all these signs.

Equally striking is the evidence provided by the case-
histories in this and the other books of *Epidemics* showing how
some Greek doctors put these principles into practice. The
case-histories (three groups of which are included in our
selection, pp. 102–12, 113–21 and 127–38) contain day by day
records of the progress of individual patients' conditions. The
entry under each day varies from a single remark to a lengthy
description of some nine or ten lines, and in some cases occasional observations continue to be recorded up to the 120th

day from the onset of the disease. The primary aim of these case-histories is to provide as exact a record as possible of the cases investigated: they contain few interpretative comments and no overall theory of diseases is stated in them. Yet the terms used in the descriptions are, of course, in many cases 'theory-laden', and they reveal certain assumptions concerning the nature and causes of diseases. Thus although *Epidemics* I and III present no schematic doctrine of humours – such as we have in *The Nature of Man*, for example – they often refer to 'bilious' and 'phlegmatic' matter in the patients' discharges. Again they adopt the common Greek view that the course of 'acute' diseases, especially, is determined by 'critical days', when marked changes take place in the patient's symptoms, and this doctrine of critical days must be recognized as one of the main motives for carrying out and recording *daily* observations. The 'constitutions' that accompany the case-histories suggest general conclusions concerning the periodicities of crises, notably concerning the regular occurrence of paroxysms and crises on either the odd, or the even,* days as numbered from the outset of the disease. Indeed in some places (for example *Epidemics* I, ch. 26) there are elaborate schemata concerning these periodicities, although elsewhere the difficulty of generalization and the variety in the courses and outcomes of diseases are remarked. The case-histories are models of painstaking and systematic observation. But these descriptions clearly reflect the writers' theoretical assumptions and interests, and not surprisingly modern clinicians have, in many cases, found it impossible to identify, on the basis of these accounts, what the patients were suffering from.

---

* The idea of the importance of the distinction between odd and even days has, in one form or another, proved exceptionally long-lived. My father tells me that as a medical student at St Bartholemew's in the early 1920s he was taught to watch for the crisis in pneumonia on the uneven days, especially the seventh, and in the sixteenth edition of William Osler's *The Principles and Practice of Medicine* (ed. H. A. Christian, New York and London, 1947, p. 49) it is remarked that 'from the time of Hippocrates it [the crisis in pneumonia] has been thought to be more frequent on the uneven days, particularly the fifth and seventh; the latter has the largest number of cases (Musser and Norris).'

Even more than their views on the conduct of a clinical examination, the theories that the Greek doctors adopted concerning treatment and cure are influenced by their pathological doctrines. The commonest theory, derived no doubt from popular beliefs but expressed as a general doctrine in several Hippocratic texts, is that opposites are a cure for opposites. Thus *Breaths* states that proposition baldly in its opening chapter. *Places in Man* (ch. 42) says that 'pains are cured by their opposites'. And *The Nature of Man* (ch. 9) puts it as follows: 'Diseases caused by overeating are cured by fasting; those caused by starvation are cured by feeding up. Diseases caused by exertion are cured by rest; those caused by indolence are cured by exertion. To put it briefly; the physician should treat disease by the principle of opposition to the cause of the disease according to its form, its seasonal and age incidence, countering tenseness by relaxation and *vice versa*. This will bring the patient most relief and seems to me to be the principle of healing.'

Elsewhere, however, doubts and criticisms are expressed concerning the application of certain theories of this type in practice. The writer of *Tradition in Medicine* attacks those who based their theories on hot, cold, dry and wet particularly on this score. In chapter 13 he suggests that they should consider the case of a man of weak constitution who has fallen sick through eating unsuitable food such as uncooked wheat and meat. 'What remedy, then,' he asks, 'should be employed for someone in this condition? Heat or cold or dryness or wetness? It must obviously be one of them because these are the causes of disease, and the remedy lies in the application of the opposite principle according to their theory.' Later (ch. 15) he has this to say of his opponents: 'Rather, I fancy, the diets they prescribe are exactly the same as those we all employ, but they impute heat to one substance, cold to another, dryness to a third and wetness to a fourth. It would be useless to bid a sick man to "take something hot". He would immediately ask "What?" Whereupon the doctor must either talk some technical gibberish or take refuge in some known solid substance.' These are forceful criticisms. Yet, as is clear from the

writer's own use of the pair 'overeating' and 'undernourishment' in chapters 9 and 10, his objection is not so much to the
principle that opposites should be countered by opposites, as
to the particular opposites, 'hot', 'cold', 'dry' and 'wet', that
his opponents referred to.

As these examples show, some theories of treatment were
highly schematic and dogmatic. Once again, however, we
must do justice to the variety of different points of view
represented among the Hippocratic writers. While some
treatises, such as *The Science of Medicine* (e.g. ch. 11), confidently proclaim that all diseases are in principle curable, other
works adopt a much more cautious and pragmatic approach.
The difficulty of effecting a cure, indeed the helplessness of
the doctor, are often expressed. In the case-histories in
Epidemics III, for instance, we read: 'It was impossible to do
anything to help her; she died' (case 9, first series); and again:
'No treatment which he received did him any good' (case 5,
second series); and in the 'constitution' in the same book
(ch. 8): 'This condition responded only with difficulty to
medicines, and in most cases purgatives did additional harm.'
The philosophy of *Epidemics* I is expressed in chapter 11:
'Practise two things in your dealings with disease: either help
or do not harm the patient.'

The dangers of doing positive harm to the patient with the
wrong treatment are frequently referred to. Thus *Regimen in
Acute Diseases* (ch. 26) notes with disapproval the theory of
some doctors that when a violent change takes place in the
body, it must be countered with another violent change: 'I
know that physicians do the exact opposite of what is correct.'
From the *Aphorisms* (Sec. VI, ch. 38) we may note: 'It is
better not to treat those who have internal cancers since, if
treated, they die quickly; but if not treated they last a long
time.' The surgical treatises, especially, repeatedly criticize
incorrect or harmful treatment. From among some two dozen
such passages, we may mention two. Chapter 25 of *Fractures*
comments as follows on incorrect bandaging in fractures:
'Then there are others who treat such cases at once with
bandages, applying them on either side, while they leave a

vacancy at the wound itself and let it be exposed. Afterwards, they put one of the cleansing applications on the wound, and treat it with pads steeped in wine, or with crude wool. This treatment is bad, and those who use it probably show the greatest folly in their treatment of other fractures as well as these . . . ' And the writer is especially indignant that, despite the harmful effects of such treatment, the doctors in question persist in treating other cases in the same way: 'They finally take off the dressings, when they find there is aggravation, and treat it for the future without bandaging. Yet none the less, if they get another wound of the same sort, they use the same treatment, for they do not suppose that the outside bandaging and exposure of the wound is to blame, but some mishap.' The other major surgical treatise, *Joints*, has this to say concerning the harmful effects of treating dislocation of the shoulder incorrectly with cauterization: 'I know of no one who uses the correct treatment, some not even attempting to take it in hand, while others have theories and practices the reverse of what is appropriate. For many practitioners cauterize shoulders liable to dislocation at the top and in front where the head of the humerus forms a prominence, and behind a little away from the top of the shoulder . . . Now these cauterizations rather bring it about than prevent it, for they shut out the head of the humerus from the space above it' (ch. 11).

In practice, the methods of treatment mentioned in the Hippocratic Corpus consist of a very few general types. The most important are surgery (especially the treatment of fractures and dislocations, but including also the use of the knife, trephining and cautery), blood-letting, the administration of purges, emetics and suppositories, baths, fomentations, ointments and plasters, and, especially, the control of regimen – diet and exercise.* The Greek doctors had, of course, no

* These general methods as such were not original to the Hippocratic authors, and certainly so far as the use of drugs, ointments and plasters goes, these remained on the whole comparatively simple: we have nothing in the Hippocratic Corpus to set beside the complex preparations described at length in, for example, Galen. On the other hand certain developments did, it seems, take place in the fifth and fourth centuries in both surgery and dietetics. In the latter, especially, the Hippocratic

antibiotics, and no reliable anaesthetic or antiseptic agents. Ideas of nursing and of hygiene were primitive. Most patients – and all bed-ridden cases – were treated in their own homes, cared for by their relatives. Although in later antiquity we can find precursors to the modern hospital both in the Roman military *valetudinaria* or sick-bays and in civilian institutions of the early Byzantine period, such as that founded by St Basil at Caesarea in the mid fourth century A.D., nothing of the kind existed in the time of the Hippocratic writers.

The author of *Decorum* recommends both that the doctor should make frequent visits, and that he should leave one of his pupils, not a layman, in charge of the patient in his absence, but how well these recommendations were carried out, we do not know. Again, although baths were occasionally prescribed, this was as part of a particular course of treatment, not as a routine measure of hygiene, and in general the importance of cleanliness was little appreciated. In the treatise that describes the doctor's surgery or consulting-room (*The Surgery*), more attention is paid to the room being well lit than to its being clean. Indeed although the need for bandages to be clean is noted, the need for the room itself to be kept clean is not, although the writer covers such other detailed points as the correct length of the surgeon's fingernails.

The recommendations not to harm the patient must, then, be understood against this background. Few remedies were available, and some of these were drastic ones. We have no reason to disbelieve the Hippocratic doctors when they observe that many commonly used treatments did more harm than good. This applies to blood-letting, cautery, the pre-

---

writers sometimes worked out extremely elaborate regimens for the healthy as well as for the sick. The surgical treatises refer to the use of some complex apparatus, especially for reducing dislocations and including the famous Hippocratic Bench, described at *Joints* chs. 72–3 (cf. *Fractures*, ch. 13), but the attitudes of the authors are ambivalent. Thus the writer of *Joints* praises those who invent devices that are 'in accordance with nature' (ch. 42), but he criticizes his colleagues for using marvellous methods in order to impress their clientèle (chs. 42 and 44) and he insists that where two methods achieve the same result, the simpler is to be preferred (ch. 78).

scription of a sudden change of diet and the use of potent drugs such as black hellebore (the dangers of which are remarked, for instance, in the *Aphorisms*, Sec. IV, ch. 16). In the circumstances, caution and a pragmatic approach were clearly in order: the doctor started with a mild treatment and continued or modified it according to the patient's response. For some conditions, Hippocratic treatments were, we may believe, effective enough. Thus the account of how to treat dislocations and fractures in the surgical treatises is, on the whole, very sound. But for many other complaints, and particularly for the acute diseases to which they devote so much of their attention, the Hippocratic doctors could do little more than let nature take its course, keeping the patient as comfortable as possible and doing nothing to exacerbate his condition, but with little hope that the diet or drugs prescribed would bring about a cure.

MEDICINE, EDUCATION AND SCIENCE

The technical or 'professional' training of the Hippocratic doctor was both far less extensive and less clearly defined than that of his modern counterpart. Conversely it was easier and commoner for men who had no practical experience of medicine to interest themselves in matters that would nowadays be deemed the province of the qualified doctor. Educated laymen and experts, when they could be distinguished at all, certainly had no problems in communicating with one another. From the fifth century B.C. onwards, there is evidence of a quite widespread interest in various aspects of medicine and medical theory outside the circles of those who actually practised as doctors. This may be seen as part of the more general phenomenon of the expansion of both 'higher' and 'technical' education in that period, associated with, but not confined to, the rise of the professional teacher or sophist. Several of the Hippocratic treatises are, as we noted, lectures addressed to a general audience. Indeed some, such as the work called *Breaths* and perhaps also *The Science of Medicine*, were composed by men who were professional lecturers, rather than professional doctors, if one may distinguish between these two activities while recognizing that many individuals engaged in both.

A remarkable passage in *The Nature of Man* shows that even quite technical subjects were lectured on in public to a general audience. This treatise begins by explaining that 'this lecture is not intended for those who are accustomed to hear discourses which inquire more deeply into the human constitution than is profitable for medical study. I am not going to assert that man is all air, or fire, or water, or earth . . .' The author is outspoken in his criticisms of those who indulge in such theorizing: they do not know what they are talking about. 'A good illustration of this is provided by attending their disputations when the same disputants are present and the same audience; the same man never wins the argument three times running, it is first one and then the other and sometimes the one who happens to have the glibbest tongue.' The implication is that the question of the ultimate constituents of the human body was, at this time, the subject of competitive public debates in which several speakers participated and the audience itself decided who was the winner.

Further evidence of the general interest in certain aspects of medicine outside the doctors comes from Aristotle. In the *Politics* (1282a 3ff.) he distinguishes between three kinds of persons who have a claim to speak on medical matters: first the ordinary practitioner, second the 'master-craftsman', and third the man 'who is educated in the art', that is, the man who has studied medicine but does not necessarily practise it. Elsewhere Aristotle (himself the son of a doctor) points out the relevance of the study of health for the inquiry concerning nature in general. It is the business of the student of nature (*physikos*), he says in the *De sensu* (436a 17ff.), 'to inquire into the first principles of health and disease . . . Generally, then, most of those who study nature end by dealing with medicine, while those of the doctors who practise their art in a more philosophical manner take their medical principles from nature.' And he makes a similar point in the *De respiratione* (480b 26ff.): 'The more subtle and inquisitive doctors speak about nature and claim to derive their principles from it, while the more accomplished investigators of nature generally end by a study of the principles of medicine.'

The place that the study of medicine might occupy in the interests of others besides the doctors can be illustrated by the fact that a philosopher such as Plato gave a detailed, six-page account of the origins of diseases in his cosmological dialogue, the *Timaeus*. Admittedly many modern scholars largely or even wholly discount Plato's efforts to give an account of the physical world in that dialogue. Yet to this two remarks must be made. Firstly, the *Timaeus* must be taken seriously in this respect at least, that Plato was deeply concerned to show the element of order and design in the cosmos, which he believed to be the best possible; as he puts it at *Timaeus* 92c, it is a 'perceptible god, greatest and best and fairest and most perfect'. Secondly, whatever modern scholars may think of the account of diseases in particular – and it must be agreed to be very largely derivative – it evidently impressed some ancient commentators. In the papyrus Anonymus Londinensis, that preserves parts of Meno's history of medicine written probably at the end of the fourth century B.C., more space is devoted to Plato than to any other theorist, including Hippocrates. Even when we make due allowance first for the fragmentary nature of the papyrus itself and then for the special attention paid to anything Plato wrote, it is clear that the physiological and medical doctrines of the *Timaeus* commanded considerable respect in the fourth century. Nor should we be surprised at this, or by the space that Plato devoted to the problem of disease in the *Timaeus*, when we recall that other writers, too, whose theories are known to us from Meno or from other sources, had many other interests besides medicine. Thus the Pythagorean Philolaus of Croton, whose doctrines are reported at Anonymus Londinensis XVIII, 8ff., worked in a variety of fields, including astronomy and geometry. And in the next century Aristotle refers on several occasions to a discussion of health and disease – which shows that he at least planned such a work, although if he wrote it it has not survived.

Medicine was, then, sometimes included in the study of

nature, and as such could be, and was, investigated by men who styled themselves philosophers rather than doctors. Moreover the relation between medicine and the rest of the study of nature was a subject of dispute. Aristotle's views on this have already been noted. From the medical writers we may cite two sharply opposed opinions. The author of *Regimen* I subordinates dietetics, at least, to physiology. In chapter 2 he writes: 'I maintain that he who intends to write correctly concerning the regimen of man must first know and discern the nature of man in general, that is know from what things he is originally composed, and discern by what parts he is controlled', and later he demands that the doctor should know such matters as the risings and settings of the stars, in order to be able to guard against the 'changes and excesses' of 'foods and drinks and winds and the whole cosmos from which diseases come to men'. The position adopted in *Tradition in Medicine* is the antithesis of this. This treatise draws a fundamental distinction between medicine and such subjects as astronomy and geology in chapter 1 (see above p. 27), and in chapter 20 the writer claims not merely that medicine is independent of philosophy, but that so far from medicine being subordinate to 'physics', the reverse applies and the study of nature in general should be approached through the study of medicine.

I think I have discussed this subject sufficiently, but there are some doctors and sophists who maintain that no one can understand the science of medicine unless he knows what man is; that anyone who proposes to treat men for their illnesses must first learn of such things. Their discourse then tends to philosophy as may be seen in the writings of Empedocles and all the others who have ever written about Nature; they discuss the origins of man and of what he was created . . . I do not believe that any clear knowledge of Nature can be obtained from any source other than a study of medicine and then only through a thorough mastery of this science.

In attempting to determine the role of medicine in the development of ancient natural science we must first recall that the conceptual categories that the Greeks employed to refer to what we term 'science' and 'philosophy' are, in certain respects, quite different from our own. There was no category of 'science' as such at all, and 'philosophy' could include not only ethics and logic, but also *physikē*, a category much wider than our 'physics' since it comprised the whole of the inquiry concerning nature. In many cases the ideas and doctrines of the medical writers run parallel to those of the philosophers and form part of the same general development in the study of nature. One example that has already been mentioned concerns causation: the investigation of the factors responsible for a disease – or for an improvement in a patient's condition – helped to focus attention on the distinction between a cause and a coincidence, and several medical writers insist that every phenomenon has a cause. If at the very beginning of scientific inquiry the rejection of mythological explanations is an important, even if negative, step, one of the best extant examples of this from early Greek science is the treatise *The Sacred Disease*, which brings a powerful set of arguments to bear to refute the view that the sacred disease is subject to divine intervention. There can be no question, the writer argues, of the sacred disease *not* having a natural cause just like every other disease, and he offers his own explanation of its origin, namely that it arises from a phlegmatic discharge that blocks the vessels communicating with the brain. Indeed taking the war into his enemies' camp he maintains that it is positively impious to suggest that any divine agency would be responsible for this or any other sickness. As for the remedies that the quacks used, 'purifications', 'sanctifications' and the like, he dismisses these as so much chicanery: 'They also employ other pretexts', he says (ch. 2), 'so that, if the patient be cured, their reputation for cleverness is enhanced while, if he dies, they can excuse themselves by explaining that the gods are to blame while they themselves did nothing wrong.'

But if the work of the doctors was often solidary with that of the philosophers, the opposition between *some* medical writers and *some* philosophers was also fruitful for Greek science. From the point of view of the development of Greek science as a whole, those who wrote chiefly on medical subjects made fundamental contributions in three main areas especially. Firstly, in the great epistemological debate on the foundations of knowledge and the relative trustworthiness of reason and sensation, some of the medical writers not only insist on the importance of observation in principle, but also provide, in practice, some remarkable examples of sustained and meticulous observation. Secondly, there is the growing recognition, during the fifth and fourth centuries, of the distinctions between different intellectual disciplines and, connected with this, a growing awareness of methodological issues. Here too important contributions are made by medical writers, particularly by the author of *Tradition in Medicine*. Thirdly, in the growing discussion of the relative importance of theoretical knowledge and its practical applications, some of the medical writers – again in contrast to some philosophers – stood firmly for the principle that what counted in medicine was the latter.

The first point concerns the use of observation. Early Greek science as a whole has often been criticized as excessively rational and dogmatic, and the criticism has a good deal of force. Certainly in the debate between reason and sensation some of the philosophers not only argued that reason is to be preferred, but also tended positively to denigrate sensation. The first to do so was Parmenides in the early fifth century, and this tradition finds its most notable exponent in Plato, although immediately after him Aristotle goes out of his way, particularly in his biological works, to defend and support the practice of observation. Dogmatism is undoubtedly a marked feature of much of Greek medicine as well as of early Greek philosophy. Yet it is in medicine that we find the best early examples of systematic and meticulous observation. As we have noted, several works insist that prognosis must be based on a very thorough examination of the patient, and the case-histories in the *Epidemics* show that some of the doctors were capable of carrying out detailed and sustained observations.

Secondly, we have seen that in one text especially, *Tradition in Medicine*, methods are explicitly discussed at some length, the two most important ideas being (1) the recognition of the inexactness of medicine, and (2) the rejection of theories based on untestable 'postulates' or 'hypotheses'. Putting these points now into the wider perspective of the development of Greek natural science as a whole, we may note that they represent the first attempt to establish distinctions between the methods of different inquiries. The early history of the use of hypotheses is obscure and controversial. The Greek term *hypothesis* itself is used only rarely in the sense of 'assumption' or 'hypothesis' before Plato. Indeed when Plato first introduces it in the *Meno* (86e ff.), he feels it necessary to explain it by referring to the practice of geometers. In geometry, a proof is undertaken by means of a 'hypothesis' when it is first agreed that if a certain condition is fulfilled, the required conclusion follows, whereas if not, the conclusion does not: and attention is then devoted to considering whether the condition in question is satisfied. 'Hypothesis' had, then, it seems, a fairly technical use in the context of mathematics (where it resembles the argument later known as analysis), although the lack of first-hand evidence about Greek mathematics before Plato precludes our saying precisely when that use was first developed, and it is not clear in particular whether it was known to the author of *Tradition in Medicine*.

What is remarkable, however, is that *Tradition in Medicine* already attacks the use of 'hypotheses' in general. He mentions their applications to obscure and problematic subjects such as astronomy and geology, but it is not the case that he approves of their use even there. On the contrary, the fact that those studies have to make use of some hypothesis is enough to condemn them in his eyes. His objection is a general one, to the use of any unwarranted assumption, any assumption for which there is, in his own words, 'no sure criterion'. We have commented on the apparent discrepancy between the writer's own practice in the physiological and pathological doctrines he puts forward, and the methodological recommendations in chapter 1. But that does not diminish the importance of those recommendations as the first, admittedly imprecise, statement that arbitrary assumptions should be excluded and that theories should be testable.

Furthermore, while this writer makes explicit a demand that

there should be a criterion, other Hippocratic authors too refer to how particular ideas work out in practice, if not in the context of their general physical doctrines, at least in that of theories of treatment. Again Greek science as a whole has been criticized – with some justification – for the lack of attention paid to practical applications of theoretical ideas. Certainly among the philosophers we find both Plato (e.g. *Republic* 527d ff.) and Aristotle (e.g. *Metaphysics* 981b 17ff.) arguing for the superiority of theoretical to practical branches of knowledge. But again some of the medical writers provide something of an exception by their evident interest in the practical consequences of different medical ideas in the actual day to day treatment of the sick.

The dominant concerns of the medical writers were those of establishing the causes of diseases, of prognosis and of treatment. Most of their main physical doctrines were, as we have seen, heavily influenced by those of the philosophers; the knowledge of physiological processes, and even of basic anatomy, in the medical writers was often rudimentary. Thus digestion, like disease, was often viewed in terms of a struggle between opposed powers, in which the stomach had to overcome the food by cooking or 'concocting' it, or the food was thought of as absorbed into the body by a process of the assimilation of 'like' substances to their like. Again ideas about the movement of different substances in the body were generally extremely vague. Thus *The Sacred Disease* represents air, which the author believes is responsible for consciousness and intelligence, as being in constant movement through the body. The air we breathe in through the mouth and nose passes first to the brain and 'then the greater part goes to the stomach' (to cool it), 'but some flows into the lungs and blood-vessels' (ch. 10); and, as noted, the author explains the cause of the sacred disease as the obstruction, by phlegm, of the air in these vessels. The accounts the Hippocratic writers give of the main blood-vessels in the body are highly schematic and are often directly related to, when not simply derived from, current practices in venesection. Thus *The Nature of Man* (ch. 11) contains a description of four main pairs of blood-vessels, the first of which 'runs from the back of the head, through the neck, and, weaving its way externally along the spine, passes into the legs, traverses the calves and the outer aspect of the ankle, and reaches the feet'. Whereupon the writer adds: 'Venesection for pains in the back and loins should therefore be practised in the hollow of the knee or externally at the ankle.' The vagueness of early Greek ideas on this subject is reflected in their terminology. Although certain Hippocratic works draw some distinction between *phlebes* and *artēriai*, these terms do not coincide with our 'veins' and 'arteries'. *Artēria* was regularly used, for example, of the principal ducts of the respiratory tract, the windpipe and the bronchi (our term trachea comes from the Greek name for that duct *hē tracheia artēria*, literally 'the rough artery'); *phlebes*, while generally used of blood-vessels, whether veins or arteries, was not confined to vessels that carry blood and/or air.

One special area which was the subject of more concentrated inquiries and speculation was the problem of generation, including the origin of sex differentiation and heredity. The problems are alluded to in several treatises, but one group in particular attempts a more systematic discussion. These are the embryological treatises, *The Seed*, *The Nature of the Child* and *Diseases* IV (the first two are included in this selection), which were almost certainly by the same writer and may have been originally composed as a unity. Many of the basic ideas in these three works are not original. Thus the writer puts forward a version of the four-humour theory in chapter 3, where he identifies four kinds of innate bodily fluid, although where *The Nature of Man* for instance gives black bile as the fourth humour, along with bile, blood and phlegm, *The Seed* gives water. The writer also attaches importance to the process of the assimilation of 'like' substances to one another, and this too was a common idea. Nor is the fundamental notion that underlies his explanation of heredity likely to be original. He holds that the seed is drawn from every part of the body: it contains something of, and so can reproduce, every part. But this theory, the so-called 'pangenesis' doctrine, was very probably first put forward by the fifth-century atomist philosopher Democritus.

Nevertheless two features of the treatment of the problem of generation in the embryological treatises are of particular interest, first the attempt at empirical research, and second the use of analogies. In chapter 29 of *The Nature of the Child* we have the first extant reference to detailed observations of the growth of an embryo chick. 'If you take twenty or more eggs and place them to hatch under two or more fowls, and on each day, starting from the second right up until the day on which the egg is hatched, you take one egg, break it open and examine it, you will find that everything is as I have described – making allowance of course for the degree to which one can compare the growth of a chicken with that of a human being.' The account is extremely brief (when Aristotle undertook a similar inquiry in the *Historia Animalium*, VI, ch. 3, his description was much fuller) and how much the writer learned from his observations is hard to say – he uses them merely to support the view that 'the seed is contained in a membrane which has an umbilicus in the centre' and to show that there are membranes extending from the umbilicus. Nevertheless the important point is that the idea of conducting such an investigation had occurred. Once again we should remember, in the background, the controversy on the relative merits of reason and sensation. While many ancient embryologists were content to arrive at their theories chiefly or even solely by the use of abstract argument, this author, at least, recognized the need not merely to open an egg to see what he could see, but to conduct a comparatively sustained set of observations.

In his attempt to tackle the major problems connected with generation, the writer's main tool is analogy. The three treatises contain a wealth of comparisons of varying complexity. One of the simpler ones is the comparison in *The Nature of the Child* (ch. 12) between the formation of a membrane round the seed in the womb and the formation of a membraneous surface on bread as it is cooked. Two types of analogies are particularly common, those with mechanical processes and those with plants. In chapter 17 of *The Nature of the Child*, the author illustrates his doctrine that like substances go to like, under the action of breath or air, by comparison with the behaviour of earth, sand and lead filings when placed in a bladder and covered with water. If you insert a pipe into the bladder and blow into it, he claims, 'first of all the ingredients will be thoroughly mixed up with the water, but after you have blown for a time, the lead will move towards the lead, the sand towards the sand, and the earth towards the earth. Now allow the ingredients to dry out and examine them by cutting around the bladder: you will find that like ingredients have gone to join like.' *Diseases* IV (ch. 39) provides another example where the writer illustrates a highly conjectural theory concerning a physiological process by citing an analogy with a comparatively simple mechanical one. There he describes setting up an arrangement of three or more vessels on a piece of level ground. The vessels are joined with communicating pipes and the writer remarks that if water is poured into one of the vessels, it finds the same level in the others, and that the system as a whole can be filled or emptied by filling or emptying one of the vessels. But this comparison is then used to suggest that 'the same thing happens in the body'. The writer maintains that 'in the same way' the humours travel between the different 'sources' in the body, that is the stomach, heart, head, spleen and liver.

Plant analogies too serve a similar role in suggesting, and supporting, theories on a variety of obscure topics. *The Seed* (ch. 9) suggests a comparison with the way in which, when a cucumber is grown in a vessel, the shape and size of the cucumber are determined by those of the vessel – in order to support the theory that the shape and size of the embryo are influenced by those of the womb. In the next chapter the writer further illustrates his view that the embryo may become deformed when constricted in the womb by comparison with a tree: 'A similar thing happens to trees which have insufficient space in the earth, being obstructed by a stone or the like. They grow up twisted, or thick in some places and slender in others, and this is what happens to the child as well, if one part of the

47

womb constricts some part of its body more than another.'
Then in *The Nature of the Child* (chs. 22–27) there is a long
digression on the growth of plants which is used to suggest
that the womb stands in the same relation to the embryo as the
earth to the plants that grow in it, and which ends (ch. 27):
'Now it is in just the same way that the child in the womb
lives from its mother, and it is on the condition of health of
the mother that the condition of health of the child depends.
But in fact, if you review what I have said, you will find that
from beginning to end the process of growth in plants and in
humans is exactly the same.'

Analogies were employed extensively from the very begin-
nings of Greek science,* and during the fifth and fourth
centuries their use develops, becoming both more elaborate
and more self-conscious. The philosopher Anaxagoras, for
instance, stated the principle that 'phenomena' – that is,
things we can see – should be used as a 'vision' of 'what is
unclear', and although this dictum covers more than the use of
analogy, it certainly includes it. Although the embryological
treatises do not refer explicitly to Anaxagoras, the writer
would no doubt have agreed with him and approved his
principle as a way of investigating the hidden internal function-
ings of the body. The frequency with which analogies of
different kinds are proposed, and the elaborate character of
those that refer not just to well-known facts but to observations of the behaviour of substances in special, artificial conditions, are remarkable and suggest a deliberate method of
procedure on the writer's part. Of course the method of
analogy suffers from serious limitations. First there is the danger of mistaking an analogy for a demonstration. Second there
is the temptation to ignore the negative analogy – the points
of difference between the things compared – though we find
some recognition in the embryological treatises that certain
analogies, such as that between the chick and the human
embryo, are not exact. And conversely the positive analogy
(or points of similarity) may be, and often is, overestimated:

* Their use is discussed in part 2 of my *Polarity and Analogy*, Cambridge,
1966.

sometimes, indeed, when the comparison is between what is known and what is completely unknown, the suggestion that there is some similarity between the two may be a matter of pure conjecture. Nevertheless the method can be, and often was, extremely fruitful in suggesting, if not in establishing, hypotheses. In the investigation of vital processes within the body it was often impracticable or quite impossible to carry out direct observations, let alone experiments. In these circumstances, drawing a comparison with what can be observed outside the body provides a means of bringing empirical data to bear on the problem, and in some cases in the embryological treatises the comparisons involved conducting simple tests on substances outside the body. The fact that analogies were often used uncritically (a feature that is not confined, of course, to their use by the ancient Greeks) should not cause us to underestimate their importance as a method of discovery, especially in the investigation of problems beyond the reach of direct observation.

Finally, having noted the comparative lack of interest in internal anatomy shown by most Hippocratic writers, we must mention the one treatise that is the exception to this general rule. This is *The Heart*, a work which is also exceptional in its date: as already noted, it may well belong to the third century. Parts of the work are very obscure and the writer adopts some extremely speculative doctrines, such as the notion that man's intelligence resides in the left chamber of the heart. Yet he gives a quite detailed account of the anatomy of the heart. He describes it as a 'strong muscle' and distinguishes the 'ears' – that is the auricles and atria – noting in chapter 8 that they do not contract simultaneously with the ventricles but have a separate movement of their own. Then in chapters 10–12 he describes the 'hidden membranes' of the heart, identifying the semi-lunar valves at the base of the aorta and the pulmonary artery and perhaps also (though here text and interpretation are more problematic) the atrio-ventricular valves. Thus on the semi-lunar valves he writes (ch. 10): 'Now there is a pair of these arteries, and on the entrance of each three membranes have been contrived,

49

with their edges rounded to the approximate extent of a semicircle. When they come together it is wonderful to see how precisely they close off the entrance to the arteries.'

Unlike most of the Hippocratic treatises, which make little or no use of dissection, the account of the heart in this work clearly depends on it. The method was first used extensively on animals by Aristotle, and although the point is disputed, there seems no good reason to deny what Celsus (among others) reports, namely that the Hellenistic biologists Herophilus and Erasistratus dissected human beings – although how extensive their dissections were is, to be sure, a matter of doubt. The works of Herophilus and Erasistratus themselves are lost, and we have to rely on quotations preserved in such later writers as Rufus and Galen. Whether or not *The Heart* is directly influenced by Erasistratus (who is credited by Galen with the discovery of the valves of the heart), it provides valuable direct evidence concerning the advances in anatomical knowledge that stemmed from the use of dissection in the third century. At the same time, we cannot but be struck by the contrast between this work and the rest of the Hippocratic treatises, where an interest in anatomy was generally confined to matters directly related to treatment.

Few, if any, of our anonymous Hippocratic authors can be considered great original scientists. Yet the contributions these treatises made to the development of Greek natural science are far from negligible. In particular, they provide important evidence concerning a strain of Greek science that, while owing much to the philosophers, also criticized and opposed them, whether implicitly or explicitly, both at the level of individual theories and on the questions of general methods and aims. Although we find plenty of schematic speculation in the medical writers, too, they differed from the philosophers in this respect especially, that they combined theoretical with practical interests: their chief preoccupation was with diagnosis and treatment and their primary aim was to succeed in combating disease and restoring or preserving health. They were conscious, and proud, of the fact that medicine was a *technē*, an art, which could be judged by results and in which there were – as many writers were keen to point out – both good and bad practitioners. Their knowledge of physiology and anatomy was generally rudimentary and in many cases their ideas were soon outdated. But in other areas the work they did set a high standard: the two outstanding examples are the account of the treatment of fractures and dislocations in the surgical treatises and the case-histories in the *Epidemics*. Indeed in the latter case the standard of detailed and meticulous observation remained unsurpassed not only throughout antiquity but down to the sixteenth century.

HIPPOCRATES' REPUTATION AND INFLUENCE

The history of the reputation of Hippocrates is, from the early third century B.C., very largely one of the development of, and changes in, an ideal. Once a body of medical writings had been collected together and associated with the name of the man who had been the subject of such complimentary references in both Plato and Aristotle, Hippocrates' fame inevitably eclipsed that of all other early doctors, and he came to stand for whatever any given writer held to be most valuable in early medicine. He attracted much of the same sort of attention that was paid, at various times in antiquity, to such other great names as Pythagoras and Plato himself. More or less fictional, and more or less hagiographical, lives of Hippocrates were written, such as that which passes under the name of Soranus – although whether it was actually written by Soranus of Ephesus, who worked in the early second century A.D. and was the author of a fine treatise on gynaecology, has been disputed. Letters too were forged that purported to have passed between Hippocrates and his great contemporary, the atomist philosopher Democritus. As in modern times, so too in the ancient world, scholars who recognized that there was a Hippocratic question – and there were plenty of them from the second century B.C. – tended nevertheless to assume that Hippocrates himself must have been the author of those treatises that they most admired.

Yet we can and should distinguish both between different

phases in the growth of Hippocrates' reputation and between the different pictures of Hippocrates that were presented. Very broadly speaking, the attitude towards the past became increasingly deferential, not to say reverential, in the ancient world, especially after the second century A.D. Yet although the tendency to idealize Hippocrates generally grew in strength, the Hippocratic writings were also the subject of critical comments. Thus it is clear from the quotations of Herophilus and Erasistratus in Galen that they implicitly, and sometimes explicitly, modified and corrected Hippocratic conceptions on a variety of points. In the first and second centuries A.D. the extant writings of Rufus and Soranus establish that they were far from accepting everything that they ascribe to Hippocrates, Soranus in particular being often highly critical. To cite a single example (and many more could be given from what is preserved of his *Acute Diseases* and *Chronic Diseases* in the Latin version of Caelius Aurelianus), Soranus refers explicitly to Hippocrates at *Gynaecology* I, 45 (*CMG* IV, 31, 26ff.) in mentioning the doctrine that we find in *Aphorisms*, Sec.V, ch. 48, that 'a male foetus inclines to the right' of the womb, and 'a female to the left', and he goes on to state that this doctrine is mistaken.

Secondly one must distinguish between the idealization of Hippocrates as a doctor and the idealization of him as a medical and biological theorist. It is one thing to represent him as the epitome of a skilful, dedicated and upright medical practitioner: it is another to accept his views on problems of pathology, physiology, anatomy and physics. The two often went together, but it was only from the second century A.D. that they tended to do so regularly.

The attitude of Galen (born about A.D. 129) in many ways marks a turning point. Galen himself very soon came to be accepted as a – or even the – chief authority on medical and biological subjects, and his views on Hippocrates were correspondingly highly influential. The tradition of learned commentaries on Hippocratic texts goes back to the third century B.C.: Galen himself, according to his own account in

the treatise *On His Own Books*, composed commentaries, sometimes in several books, on no fewer than nineteen Hippocratic texts. His work on some thirteen of them, including
Epidemics I, II, III and VI, *Prognosis, Regimen in Acute Diseases,
Aphorisms, The Nature of Man, A Regimen for Health, Fractures*
and *Joints*, has survived, as also has his *Glossary* of Hippocratic
terms. When we add that he also composed a work on *The
Elements according to Hippocrates* and a major treatise on *The
Opinions of Hippocrates and Plato* (both of which are extant),
the body of Galen's writings directly or indirectly related to
Hippocrates is seen to be very considerable, making up more
than a quarter of all the work of his that has survived in Greek
– and this takes no account of the frequent occasions on which
he refers to Hippocrates in other works, usually to cite him as
an authority for the view that Galen himself upholds.

Galen was fully aware of, and frequently refers to, scholarly
disputes over the authorship of the treatises ascribed to
Hippocrates. Thus he remarks concerning *Regimen* and *A
Regimen for Health* that these works had been attributed to such
men as Euryphon, Ariston and Philistion. In his commentary
on *The Nature of Man* he notes that some saw Polybus as the
author of the work. In that case, however, he believes that
the authenticity of the treatise is proved by the evidence in
Plato's *Phaedrus*. Few would now go along with him, but
Galen held that Plato must have had *The Nature of Man* in
mind when he makes Socrates attribute to 'Hippocrates and
the true account' a method that first considers whether a thing
is simple or complex, and then, if it is complex, enumerates its
parts and their capacities (*Phaedrus* 270c ff.). Yet we should note
that it is Galen's view that even if *The Nature of Man* were by
Polybus, it would still be good evidence for the doctrines of
Hippocrates himself. Galen states that Polybus was the pupil,
as well as the son-in-law, of Hippocrates, that he took over from
Hippocrates the task of teaching the young, and most important, that he appeared not to have modified any of the doctrines
of Hippocrates in his own writings. Galen himself holds that
most of the works that were ascribed to Hippocrates were

indeed authentic, but he also believes that others that may not be from Hippocrates' own hand may still faithfully record his views.

Galen acknowledges, on occasion, that there are gaps in Hippocrates' knowledge and that there are points on which he is unclear or even mistaken. In general, however, Galen follows what he represents as the Hippocratic teaching, and treats Hippocrates not only as a distinguished practitioner, but as an authority whose views on medical and biological matters were to be adopted wherever possible. From the point of view of Hippocrates' subsequent reputation, what was crucial was that Galen saw him as the originator of the physical and physiological theory based on the four primary opposites (hot, cold, dry and wet) and the four primary simple bodies or elements (earth, water, air and fire). This was the theory that – with modifications and additions – Galen himself adopted as the foundation of his own account of the ultimate constituents of the body. He took over, for example, the distinction used by Aristotle, among others, between homoeomerous or homogeneous parts (such as flesh, bone and blood) and instrumental ones (such as foot and hand); he also held that each of the primary opposites exists in different grades – so that there are, for example, four degrees of heat; and he put forward a complex doctrine of vital powers or functions. But on the question of the fundamental constituents of physical objects he took over the four-element theory, versions of which he knew had been proposed by both Plato and Aristotle. However, on the basis of the evidence in *The Nature of Man* especially, Galen claimed that it was Hippocrates who had first identified the four primary elements and defined their qualities. As he puts it in his treatise *On the Natural Faculties* (I, ch. 2): 'Of all the doctors and philosophers we know, he [Hippocrates] was the first who undertook to demonstrate that there are, in all, four mutually interacting qualities, through the agency of which everything comes to be and passes away.' Galen certainly admired other aspects of Hippocratic teaching, but his enthusiastic endorsement of the schematism of *The Nature of Man* was especially influential.

After the second century A.D., two main changes may be noted: first the fact that, as already remarked, Galen himself rapidly became the chief authority on questions of anatomy, physiology and pathology; and second the increasingly important place that the commentary, digest or history came to occupy in medical writing. It would certainly be an exaggeration to say that medical and biological inquiry died out entirely after Galen, but more and more work in these fields – as also in philosophy, mathematics and other areas of scientific investigation – came to take the form of commentaries on, or even summaries or abridgements of, earlier texts. We can trace this development through the writings of such men as Oribasius (mid fourth century), Aetius of Amida, Alexander of Tralles (both sixth century), Paul of Aegina (seventh) and Theophilus Protospatharius (not before 600), all of whom, while not totally devoid of originality, relied heavily on earlier sources. Thus Oribasius, substantial parts of whose vast seventy-book encyclopedia of medical knowledge are extant, explains in the preface to the work that his plan is to produce an even more complete compendium than his first effort, which had been based on Galen alone. Now he will draw on 'all the best doctors', though Galen still ranks first among these 'on the grounds that he is supreme among all those who have written about the same subject, since he uses the most exact methods and definitions, as one who follows the Hippocratic principles and opinions'. To judge from the extant books, Galen is indeed Oribasius' principal source, but it is clear that he believes that Galen represents the Hippocratic tradition faithfully.

The main effort of most other later medical writers too was towards summarizing and systematizing medical knowledge, and as time goes on their summaries tend to become more concise. Thus Paul of Aegina refers to Oribasius' work in the proem to his own treatise, where he states that he considered it too bulky and so made a shorter and more convenient compendium of his own. Paul still refers to and quotes from Hippocrates quite freely, including treatises that had not been the subject of commentaries by Galen. But medical writers evidently came more and more to rely on what Galen had preserved. Thus when Theophilus cites a passage from Hippocrates' *Epidemics* VI in his work on the pulse he reveals where he found it by introducing the quotation with the words, 'Hippocrates speaks thus in the work on the differences of fevers'. There is no such Hippocratic treatise, and it is from Galen's work of that name that the quotation from Hippocrates comes.

After the division of the Roman Empire, the decline in learning was much more rapid and severe in the Roman West than in the Greek East. In the East, as we have seen, some knowledge of Galen and Hippocrates continued down to the seventh century and beyond, but in the West interest in scientific medicine sank to a low ebb. Some medical treatises continued to be written, such as the free Latin versions of Soranus' *Acute Diseases* and *Chronic Diseases* that were produced by Caelius Aurelianus in the fifth century. But otherwise what survived of Greek learning in the West was mostly contained in handbooks that owed their popularity mainly to the curiosities they retailed. Thus a hotchpotch of information on a variety of subjects was preserved in such works as the *Saturnalia* and *Commentary on the Dream of Scipio* of Macrobius (late fourth, early fifth century) and the *Institutions* of Cassiodorus (sixth century). But much less attention was paid in such works to medicine, which was not one of the seven 'liberal arts', than to subjects like astronomy and music. There are occasional references to Hippocrates in the *Saturnalia* – although one of the characters in that work describes medicine as 'the lowest dregs of physics'. Yet on medical matters literary authors came to supplant the original Greek medical writers. In the influential writings of Isidore of Seville (seventh century), medicine and biology are comparatively neglected. Although Hippocrates is mentioned as one of the founders of the 'logical' or 'rational' school of medicine in book IV of Isidore's *Etymologies*, the treatment of medicine there is both superficial and eclectic and we find the Old Testament, Virgil and Horace among the authorities cited.

Meanwhile in the East the preservation and resuscitation of the spirit of inquiry depended, from the eighth century, on the Arabs. After the fall of Alexandria in 642, knowledge of Greek medicine, as of other aspects of Greek learning, spread through the Arab world. In the ninth century, translations of Hippocrates and Galen were undertaken by Hunain, and the polymath Al Kindi covered the entire range of learning and wrote more than twenty treatises on medical subjects, including one specifically on Hippocratic medicine. Among the voluminous writings of Rhazes (Al Razi, died *c.* 925) were the *Al Mansuri*, which contained among other things a detailed account of human anatomy, and the so-called *Liber Continens*, the *Kitabu'l Hawi Fi't-Tibb*, a medical encyclopedia in twenty-three books. The intensive study of medicine was sustained in the tenth century by Ali-Abbas, who quotes Hippocrates in his anatomical treatise, and in the eleventh by Avicenna; and nearly all the great names in the western school of Arabic philosophy, including the greatest of all, Averroes (twelfth century), were also doctors and wrote on medical subjects. Most of these Arabic writers knew their Hippocrates as well as their Galen and saw themselves as upholding the best traditions of Greek medicine.

It was, of course, mainly through Latin translations, both of these Arabic authors and then of the Greeks themselves, that knowledge of Greek medicine, as of other parts of Greek science, was revived in the Christian West, although, as the manuscript tradition shows, some Latin versions of Greek texts survived throughout the Dark Ages. In this revival, two centres were particularly important. At Salerno, which had a flourishing medical school from about the late tenth century, Latin translations and paraphrases of Rhazes (among others) and of some Hippocratic works were made in the eleventh century by Constantine the African (whether he used the Arabic versions of Hippocrates or worked direct from Greek manuscripts is disputed). And at Toledo an important series of translations was done in the twelfth century by Gerard of Cremona and others; these included many Arabic medical treatises, for example the *Canon* of Avicenna. Hippocrates and Galen began to figure prominently in the curricula of the medical faculties of European universities from the fourteenth century, and many new Latin translations of their works were made in that and the next two centuries, while vernacular translations of the most popular Hippocratic works, such as the *Aphorisms* and *Prognosis*, had been initiated by the mid sixteenth century. Yet a number of late forgeries and spurious treatises, such as the prognostic work known as the *Secreta* or the *Liber Veritatis*, as well as a motley collection of letters, continued for long to pass for genuine Hippocrates, and his authority was invoked in support not only of the humoral theory but also of astrological doctrines.

By about the middle of the fourteenth century Galen had once again achieved the position of dominance, as the outstanding authority in anatomy and physiology, that he had enjoyed in the Greco-Roman world, and the advance of knowledge in both fields depended on going beyond him. Although Galen and the Galenists had been criticized before, Vesalius' attack on him, and more especially on them, may be mentioned as one of the most influential. His *De Humani Corporis Fabrica*, which was first published in 1543, combined an onslaught on Galen with an insistence on the need for dissection. While acknowledging Galen as 'easily the foremost among the teachers of anatomy', Vesalius castigates him for his errors and the Galenists for the blindness with which they followed him. 'So completely have all surrendered to [Galen's] authority,' he writes in his preface, 'that no doctor has been found to declare that in the anatomical books of Galen even the slightest error has ever been found, much less could now be found . . .'

For I am not unaware how the medical profession . . . are wont to be upset when in more than two hundred instances, in the conduct of the single course of anatomy I now exhibit in the schools, they see that Galen has failed to give a true description of the interrelation, use, and function of the parts of man . . . Yet they too, drawn by the love of truth, gradually abandon that attitude and, growing less emphatic, begin to put their faith in their own not ineffectual sight and powers of reason rather than in the writings of Galen.*

Yet the overthrow of Galen in medicine (paralleled by that of Aristotle in physics and that of Ptolemy in astronomy) left Hippocrates' own reputation in many respects untouched. Indeed in the sixteenth, seventeenth and eighteenth centuries there were many leading medical writers who continued to express their admiration for Hippocrates and who advocated a return to what he stood for. They included Guillaume de Baillou (born *c.* 1538) in France, who modelled his own case-histories on the Hippocratic *Epidemics*, Sydenham (born 1624) in England, and Boerhaave (born 1668) in Holland. For Sydenham, for instance, Hippocrates was an 'unrivalled historian of disease', who had 'founded the Art of Medicine on a solid and unshakeable basis', namely the principle that 'our natures are the physicians of diseases' and the method of 'the exact description of nature'.† But what these men admired in Hippocrates was not the anatomy and the physiology of the treatises, so much as two things particularly: firstly the detailed and meticulous clinical observations, and secondly the example he set of the doctor's devotion and concern for his patients, and of his uprightness and discretion in his dealings with them. For this example, especially, Hippocrates continued, and continues, to inspire.

Since the seventeenth century the rate of change in medicine has continued to accelerate. Yet the Hippocratic writings are still of great value to us today – and not merely in the context of the history of science. Their importance in that context – and for the historian of culture – has been described briefly above: they throw light not only on the origins and early development of medicine and on its place in Greek society, but also on the development of early Greek science as a whole. But these texts also retain a more general interest for the medical profession and for all concerned with the ethics of

*Translation from B. Farrington, in *Proceedings of the Royal Society of Medicine*, 25, 1932, Section of the History of Medicine, pp. 39–48.

†Preface to the third edition (1676) of *Observationes Medicae*, ed. G. A. Greenhill, London, 1844, para. 15, pp. 13f.

medicine. Issues relating to questions of medical ethics and of the ethics of scientific research are as live today as they have ever been, and it is instructive to study the Hippocratics not simply because the name of Hippocrates is so often invoked in such discussions, but because it is with the Hippocratic authors that the question of the duties of the doctor both to his patient and to society as a whole first come to be the subject of debate. They were the first to attempt to establish a code of behaviour for the medical profession and to define the doctor's obligations, first to see to it that medical knowledge is used for good and not for ill, and second not to exploit his privileged relationship with his patient in any way. Western medicine has come a long way since its Greek origins. Yet on questions of medical ethics and etiquette, not only do we owe a debt to the past – as the source of many of the unwritten assumptions, as well as of some of the explicit rules, that still govern the relationship between the doctor and patient – but we still have something to learn from it. While most of the anatomical, physiological and pathological doctrines in the Hippocratic writings have long since been superseded, the ideal of the selfless, dedicated and compassionate doctor they present has lost none of its relevance in the twentieth century.

MEDICINE

TRANSLATED BY
J. CHADWICK AND W. N. MANN

TRANSLATORS' INTRODUCTION*

Books in Ancient Greece could only be made by individual copying. They were therefore not produced in any great numbers and there was always a chance of errors creeping in each time a new copy was made. As these faulty texts were themselves copied so the errors were multiplied and propagated; at the same time readers or copyists might make corrections either by conjecture or by comparison with another copy. Hippocrates has of course suffered much in this process which has preserved the works for us. Our earliest manuscript is probably of the tenth century A.D.; that is, it was written at least thirteen hundred years after the works it contained were composed. We have others of the eleventh and twelfth centuries and a number of less valuable later ones. From a comparison of these it is possible to reconstruct a fairly satisfactory text, but there will obviously be places where none of the manuscripts provides a likely reading or where they all fail. In the one case we must, if possible, restore by conjecture what was the probable text, in the other we can do nothing but signify the gap or *lacuna* by a row of asterisks. Some important chapters of *Airs, Waters, Places* are lost in this way. However, the researches of generations of scholars have succeeded in restoring what is on the whole a sound and reliable text. In translation we have done our best to avoid any discussion of dubious readings, selecting where possible that which seems to make the best sense. In one or two places we have varied the traditional order to improve the connection of thought. In a very few instances we have ventured to make new suggestions which we think are demanded by the sense; but in the main we have followed the best texts available.†

*The following is an abbreviated version of the translators' original introduction.

†See below p. 354.

The numbers of paragraphs and sections are traditional and have been inserted for the benefit of readers who wish to refer to the Greek texts or other translations.

The ideal translator of Hippocrates would have to be not only acquainted with Ancient Greek but also to have a wide knowledge of practical medicine. In these days such a combination must be extremely rare, and we hope we have found a satisfactory substitute in our close co-operation in the respective fields. The translation has gone through several stages. First a careful and literal rendering was made of the Greek together with notes on the meaning and alternative interpretations; this was edited, taking careful note of the medical significance of the passage, and put into current English. This revision was then checked against the Greek original, and so on, until a mutually agreed form was reached. As far as possible we have used modern medical English, with the obvious limitations. In some cases the original Greek medical term is still in current use but it is no longer, as may be seen from the context, a correct translation, for in the course of time the word may have come to bear a more limited or even a changed meaning. Thus the Greek word *noma* means a gangrenous patch and in the Hippocratic text is used to describe this condition occurring on the tonsils. However, in current medical terminology the word is used exclusively to describe gangrene of the mouth or *cancrum oris*. It is clear that to use the English word 'noma' as a translation of the Greek word *noma* falsifies the sense. We have, however, avoided using the modern name of a disease where there is no evidence that Hippocrates appreciated its morbid identity. In such cases we have left the translation as literal as possible, and in one case we have kept the Greek word. Hippocrates refers frequently to a febrile malady he terms *causus*. It is certain that he included under this diagnosis a variety of fevers common in the Levant, but which we now separate into a number of different diseases. In many cases, enteric fever is clearly described; but as the condition, *causus*, cannot be generally identified as a single disease, the term is kept in the translation whenever it appears in the original text.

There are evident pitfalls in this joint method of translation. For instance, it is very easy, particularly in textually corrupt passages, to ascribe to Hippocrates an insight into morbid processes which he did not have. But we believe we have made few such mistakes. The alternative is to leave this selection and interpretation to the reader, but that was not our object in producing this new translation. The readers for whom this book is intended may not have the time nor opportunities to weigh and scrutinize each sentence; they will expect that to have been done for them, and they have here presented the best that the translators can provide. Nor need they fear that the necessary interpretation has obscured their view of the real Hippocrates; a glance at previous translations will show that this does not affect the major part of the work, or indeed the important conclusions.

A note on the system of counting days seems necessary. It was the Greek custom to include the days at both ends, so that the third day means the day after tomorrow, and so on – a meaning familiar in terms taken from Greek medicine such as tertian or quartan. It seems best, once this has been pointed out, to leave all the numbers as they appear in the Greek text; the reader must remember that the twentieth day, for example, means what we should in ordinary speech call the nineteenth.

CHRONOLOGICAL NOTE

Hippocrates, writing 400 years before the introduction of the Julian calendar, had no convenient method for giving dates in the year. There were several different calendars in use in Greece, and they were all based upon lunar months, so that the same date would not always fall on the same day of the solar year. It was therefore common practice to use certain astronomical events as a rough method of dating. The four obvious points are the equinoxes and solstices: 21 March, 21 September, 21 June, 22 December. These are from time to time supplemented by reference to the heliacal rising or setting of certain stars and constellations. Owing to the precession of the equinoxes these are not constant, and various factors

prevent an exact calculation of the dates Hippocrates intended.
The chief of these mentioned in the text, together with their
approximate equivalents, are:

The rising of Arcturus	10 September
The rising of the Pleiads	10 May
The setting of the Pleiads	11 November
The rising of the Dog Star	17 July

THE OATH

I SWEAR by Apollo the healer, by Aesculapius, by Health and all the powers of healing, and call to witness all the gods and goddesses that I may keep this Oath and Promise to the best of my ability and judgement.

I will pay the same respect to my master in the Science as to my parents and share my life with him and pay all my debts to him. I will regard his sons as my brothers and teach them the Science, if they desire to learn it, without fee or contract. I will hand on precepts, lectures and all other learning to my sons, to those of my master and to those pupils duly apprenticed and sworn, and to none other.

I will use my power to help the sick to the best of my ability and judgement; I will abstain from harming or wronging any man by it.

I will not give a fatal draught to anyone if I am asked, nor will I suggest any such thing. Neither will I give a woman means to procure an abortion.

I will be chaste and religious in my life and in my practice.

I will not cut, even for the stone, but I will leave such procedures to the practitioners of that craft.

Whenever I go into a house, I will go to help the sick and never with the intention of doing harm or injury. I will not abuse my position to indulge in sexual contacts with the bodies of women or of men, whether they be freemen or slaves.

Whatever I see or hear, professionally or privately, which ought not to be divulged, I will keep secret and tell no one.

If, therefore, I observe this Oath and do not violate it, may I prosper both in my life and in my profession, earning good repute among all men for all time. If I transgress and forswear this Oath, may my lot be otherwise.

THE CANON

A brief note on the characteristics desirable in a student of medicine.

ALTHOUGH the art of healing is the most noble of all the arts, yet, because of the ignorance both of its professors and of their rash critics, it has at this time fallen into the least repute of them all. The chief cause for this seems to me to be that it is the only science for which states have laid down no penalties for malpractice. Ill-repute is the only punishment and this does little harm to the quacks who are compounded of nothing else. Such men resemble dumb characters on the stage who, bearing the dress and appearance of actors, yet are not so. It is the same with the physicians; there are many in name, few in fact.

For a man to be truly suited to the practice of medicine, he must be possessed of a natural disposition for it, the necessary instruction, favourable circumstances, education, industry and time. The first requisite is a natural disposition, for a reluctant student renders every effort vain. But instruction in the science is easy when the student follows a natural bent, so long as care is taken from childhood to keep him in circumstances favourable to learning and his early education has been suitable. Prolonged industry on the part of the student is necessary if instruction, firmly planted in his mind, is to bring forth good and luxuriant fruit.

The growth of plants forms an excellent parallel to the study of medicine. Our characters resemble the soil, our masters' precepts the seed; education is the sowing of the seed in season and the circumstances of teaching resemble the climatic conditions that control the growth of plants. Industrious toil and the passage of time strengthen the plant and bring it to maturity.

The man, then, who brings these qualities to the study of medicine and who has acquired an exact knowledge of the subject before he sets out to travel from city to city, must be considered a doctor not only in name but in fact. Want of skill is a poor thing to prize and treasure. It robs a man of contentment and tranquillity night and day and makes him prone to cowardice and recklessness, the one a mark of weakness, the other of ignorance. Science and opinion are two different things; science is the father of knowledge but opinion breeds ignorance.

Holy things are revealed only to holy men. Such things must not be made known to the profane until they are initiated into the mysteries of science.

TRADITION IN MEDICINE

An explanation of the empirical basis of medicine as practised about the end of the fifth century B.C. This treatise is sometimes referred to as On Ancient Medicine.

1. In all previous attempts to speak or to write about medicine, the authors have introduced certain arbitrary postulates* into their arguments, and have reduced the causes of death and the maladies that affect mankind to a narrow compass. They have supposed that there are but one or two causes: heat or cold, moisture, dryness or anything else they may fancy. From many considerations their mistake is obvious; indeed, this is proved from their own words. They are specially to be censured since they are concerned with no bogus science, but one which all employ in a matter of the greatest importance, and one of which the good professors and practitioners are held in high repute. But besides such there are both sorry practitioners and those who hold widely divergent opinions. This could not happen were medicine a bogus science to which no consideration had ever been given and in which no discoveries had been made. For if it were so, all would be equally inexperienced and ignorant, and the condition of their patients due to nothing but the law of chance. But this is not so, and the practitioners of medicine differ greatly among themselves both in theory and practice just as happens in every other science. For this reason I do not think that medicine is in need of some new postulate, dealing, for instance, with invisible or problematic substances, and about which one must have some postulate or another in order to discuss them seriously. In such matters, medicine differs from subjects like astronomy and geology, of which a man might know the truth and lecture on it without either he or his audience being able to judge whether it were the truth or not, because there is no sure criterion.

* The term translated 'postulate' is *hypothesis*, on the meaning of which see editor's introduction, p. 43.

2. Medicine has for long possessed the qualities necessary to make a science. These are a starting point and a known method according to which many valuable discoveries have been made over a long period of time. By such a method, too, the rest of the science will be discovered if anyone who is clever enough is versed in the observations of the past and makes these the starting point of his researches. If anyone should reject these and, casting them aside, endeavour to proceed by a new method and then assert that he has made a discovery, he has been and is being deceived. A discovery cannot be made thus, and the reason why such a thing is impossible I shall endeavour to show by expounding the true nature of the science. My exposition will demonstrate clearly the impossibility of making discoveries by any other method but the orthodox one.

It seems to me to be of the greatest importance that anyone speaking of the science should confine himself to matters known to the general public, since the subject of inquiry and discourse is none other than the maladies of which they themselves fall sick. Although it were no easy matter for common people to discover for themselves the nature of their own diseases and the causes why they get worse or get better, yet it is easy for them to follow when another makes the discoveries and explains the events to them. Then when a man hears about a disease he will only have to remember his own experience of it. But if anyone departs from what is popular knowledge and does not make himself intelligible to his audience, he is not being practical. For such reasons we have no need of a postulate.

3. In the first place, the science of medicine would never have been discovered nor, indeed, sought for, were there no need for it. If sick men fared just as well eating and drinking and living exactly as healthy men do, and no better on some different regimen, there would be little need for the science. But the reason why the art of medicine became necessary was because sick men did not get well on the same regimen as the healthy, any more than they do now. What is more, I am of the opinion that our present way of living and our present

diet would not have come about if it had proved adequate for a man to eat and drink the same things as an ox or a horse and all the other animals. The produce of the earth, fruits, vegetables and grass, is the food of animals on which they grow and flourish without needing other articles of diet. In the beginning I believe that man lived on such food and the modern diet is the result of many years' discovery. Such devising was necessary because, in primitive times, men often suffered terribly from their indigestible and animal-like diet, eating raw and uncooked food, difficult to digest. They suffered as men would suffer now from such a diet, being liable to violent pain and sickness and a speedy death. Certainly such ills would probably prove less serious then than now because they were accustomed to this kind of food, but even then, such illnesses would have been serious and would have carried off the majority of a weak constitution although the stronger would survive longer, just as now some people easily digest strong meats while others suffer much pain and illness from them. For this reason I believe these primitive men sought food suitable to their constitutions and discovered that which we now use. Thus, they took wheat and wetted it, winnowed it, ground it, sifted it, and then mixed it and baked it into bread, and likewise made cakes from barley. They boiled and baked and mixed and diluted the strong raw foods with the weaker ones and subjected them to many other processes, always with a view to man's nature and his capabilities. They knew that if strong food was eaten the body could not digest it and thus it would bring about pain, sickness and death, whereas the body draws nourishment and thus grows and is healthy from food it is able to digest. What fairer or more fitting name can be given to such research and discovery than that of medicine, which was founded for the health, preservation and nourishment of man and to rid him of that diet which caused pain, sickness and death?

4. It is perhaps not unreasonable to assert that this is no science, for no one can properly be called the practitioner of a science of which the facts are unknown to none and with which all are acquainted by necessity and experience. The

discoveries of medicine are of great importance and are the result of thought and skill on the part of many people. For instance, even now trainers in athletics continue to make discoveries according to the same method; they determine what men must eat and drink to gain the greatest mastery over their bodies and to achieve the maximum strength.

5. Turning now to what is generally admitted to be the science of medicine, namely, discoveries concerning the sick, which is a science in name and boasts practitioners, let us consider whether it has the same purposes and from what origins it arose. As I have already said, I do not believe anyone would ever have looked for such a science if the same regimen were equally good for the sick and the healthy. Even now some people, the barbarians and some Greeks, who have no knowledge of medicine, go on behaving when they are ill just as they do in health. They neither abstain from nor moderate the use of the things they like. Those who sought for and found the science of medicine held the same opinion as those whom I mentioned before. First of all, I imagine, they cut down the quantity without changing the quality of the food, making the sick eat very little. But when it became clear to them that such a regimen suited and helped some of the sick but not all, and that there were some even who were in such a condition that they could not digest even a very little food, then they concluded that in some cases a more easily digested food was necessary. Thus they invented gruel by mixing a little strong food with much water, so taking away its strength by dilution and cooking. For those that could not digest even gruel, they substituted liquid nourishment, taking care that this should be of moderate dilution and quantity, neither too weak nor too strong.

6. It must be clearly understood, however, that gruel is not necessarily of assistance to everyone who is sick. In some diseases it is evident that on such a diet, the fever and pains increase, the gruel serving as nourishment to the disease, but as a source of decline and sickness to the body. In such cases were dry food to be taken, barley-cakes or bread for example, even in very small quantities, the patients would become ten

times worse than they would be on a diet of gruel, simply because of the strength of the food. Again, a man who was helped by gruel but not by dry food would be worse if he ate more of the latter than if he took only a little, and even a small quantity would give him pain. In fact, it is obvious that all the causes of such pains come to the same thing; the stronger foods are the most harmful to man whether he be in health or sickness.

7. What then is the difference in intention between the man who discovered the mode of life suitable for the sick, who is called a physician and admitted to be a scientist, and him who, from the beginning, discovered the way to prepare the food we eat now instead of the former wild and animal-like diet? I can see no difference; the discovery is one and the same thing. The one sought to do away with those articles of diet which, on account of their savage and undiluted nature, the human frame could not digest, and on which it could not remain healthy; the other discovered what a sick man could not digest in view of his particular malady. What difference is there save in the appearance, and that the one is more complicated and needs more study? Indeed, one is the forerunner of the other.

8. A comparison between the diets of a sick man and a healthy one shows that the diet of a healthy man is no more harmful to a sick man than that of a wild beast to a healthy man. Suppose a man be suffering from a disease, neither something malignant nor incurable, nor yet some trifling ailment, but one nevertheless of which he is well aware. If he were to eat bread or meat or anything else which is nourishing to a healthy man, but in smaller quantities than if he were well, he would suffer pain and run some risk. Now suppose a healthy man with neither an utterly weak nor a strong constitution were to take small quantities of a diet which would give strength and nourishment to an ox or a horse, such as vetch or barley-corn, he would suffer no less pain and run no less risk than the sick man who inopportunely ate bread or barley-cake. This proves that the whole science of medicine might be discovered by research according to these principles.

9. If it were all as simple as this, that the stronger foods are harmful and the weaker good and nourishing for men both in health and sickness, the matter were an easy one. The safest course would be to keep to the weaker food. But if a man were to eat less than enough he would make as big a mistake as if he were to eat too much. Hunger is a powerful agent in the human body; it can maim, weaken and kill. Under-nourishment gives rise to many troubles and, though they are different from those produced by over-eating, they are none the less severe because they are more diverse and more specific. One aims at some criterion as to what constitutes a correct diet, but you will find neither number nor weight to determine what this is exactly, and no other criterion than bodily feeling. Thus exactness is difficult to achieve and small errors are bound to occur. I warmly commend the physician who makes small mistakes; infallibility is rarely to be seen. Most doctors seem to me to be in the position of poor navigators. In calm weather they can conceal their mistakes, but when overtaken by a mighty storm or a violent gale, it is evident to all that it is their ignorance and error which is the ruin of the ship. So it is with the sorry doctors who are the great majority. They cure men but slightly ill, in whose treatment even the biggest mistakes would have no serious consequences. Such diseases are many and much more common than the more serious ones. When doctors make mistakes over such cases, their errors are unperceived by the layman, but when they have to treat a serious and dangerous case, a mistake or lack of skill is obvious to all, and vengeance for either error is not long delayed.

10. That over-eating should cause no less sickness than excessive fasting is easily understood by reference to the healthy. Some find it better to dine but once a day and consequently make this their custom. Others, likewise, find it is better for them to have a meal both at noon and in the evening. Then there are some who adopt one or other of these habits merely because it pleases them or because of chance circumstances. On the grounds of health it matters little to most people whether they take but one meal a day or two. But there are some who, if they do not follow their usual

custom, do not escape the result and they may be stricken with a serious illness within a day. Some there are who, if they take luncheon when this practice does not agree with them, at once become both mentally and physically dull; they yawn and become drowsy and thirsty. If subsequently they should dine as well, they suffer from wind, colic and diarrhoea and, not infrequently, this has been the start of a serious illness even though they have taken no more than twice the amount of food they have been accustomed to. Similarly, a man who is accustomed to taking luncheon because he finds that this agrees with him, cannot omit the meal without suffering great weakness, fear and faintness. In addition, his eyes become sunken, the urine more yellow and warmer, the mouth bitter, and he has a sinking feeling in his stomach. He feels dizzy, despondent and incapable of exertion. Then later when he sits down to dine, food is distasteful to him and he cannot eat his customary dinner. Instead, the food causes colic and rumblings and burns the stomach; he sleeps poorly and is disturbed by violent nightmares. With such people this too has often been the start of some illness.

11. Let us consider the reason for these things. The man who is accustomed to dine only once a day suffers, in my opinion, when he takes an extra meal because he has not waited long enough since the last. His stomach has not fully benefited from the food taken on the previous day and has neither digested nor discarded it, nor calmed down again. This new food is introduced into the stomach while it is still digesting and fermenting the previous meal. Such stomachs are slow in digestion and need rest and relaxation. The man who is accustomed to a meal at midday suffers when he has to go without, because his body needs nourishment and the food taken at the previous meal has already been used up. If no fresh food be taken his body wastes through starvation, and I attribute to this the symptoms from which I described such a man to suffer. I maintain that other healthy people will suffer from these same troubles if they fast for two or three days.

12. Those constitutions which react rapidly and severely to changes in habit are, in my opinion, the weak ones. A weak

man is next to a sick man, while a sick man is made still weaker by indiscretions in his diet. In matters requiring such nicety, it is impossible for science to be infallible. There are many things in medicine which require just as careful judgement as this matter of diet, and of these I will speak later. I contend that the science of medicine must not be rejected as non-existent or ill-investigated because it may sometimes fail in exactness. Even if it is not always accurate in every respect, the fact that it is able to approach close to a standard of infallibility as a result of reasoning, where before there was great ignorance, should command respect for the discoveries of medical science. Such discoveries are the product of good and true investigation, not chance happenings.

13. I wish now to return to those whose idea of research in the science is based upon the new method: the supposition of certain postulates. They would suppose that there is some principle harmful to man: heat or cold, wetness or dryness, and that the right way to bring about cures is to correct cold with warmth, or dryness with moisture and so on. On such an assumption let us consider the case of a man of weak constitution. Suppose he eats grains of wheat as they come straight from the threshing-floor and raw meat, and suppose he drinks water. If he continues with such a diet I am well aware that he will suffer terribly. He will suffer pain and his body will become enfeebled; his stomach will be disordered and he will not be able to live long. What remedy, then, should be employed for someone in this condition? Heat or cold or dryness or wetness? It must obviously be one of them because these are the causes of disease, and the remedy lies in the application of the opposite principle according to their theory. Really, of course, the surest remedy is to stop such a diet and to give him bread instead of grains of wheat, cooked instead of raw meat and wine to drink with it. Such a change is bound to bring back health so long as this has not been completely wrecked by the prolonged consumption of his former diet. What conclusion shall we draw? That he was suffering from cold and the remedy cured him because it was hot, or the reverse of this? I think this is a question which

would greatly puzzle anyone who was asked it. What was taken away in preparing bread from wheat; heat, cold, moisture or dryness? Bread is subjected to fire and water and many other things in the course of its preparation, each of which has its own effect. Some of the original qualities of wheat are lost, some are mixed and compounded with others.

14. I know too that the body is affected differently by bread according to the manner in which it is prepared. It differs according as it is made from pure flour or meal with bran, whether it is prepared from winnowed or unwinnowed wheat, whether it is mixed with much water or little, whether well mixed or poorly mixed, over-baked or under-baked, and countless other points besides. The same is true of the preparation of barley-meal. The influence of each process is considerable and each has a totally different effect from another. How can anyone who has not considered such matters and come to understand them, possibly know anything of the diseases that afflict mankind? Each one of the substances of a man's diet acts upon his body and changes it in some way and upon these changes his whole life depends, whether he be in health, in sickness, or convalescent. To be sure, there can be little knowledge more necessary. The early investigators in this subject carried out their researches well and along the right lines. They referred everything to the nature of the human body, and they thought such a science worthy of being ascribed to a god, as is now believed. They never imagined that it was heat or cold, or wetness or dryness, which either harmed a man or was necessary to his health. They attributed disease to some factor stronger and more powerful than the human body which the body could not master. It was such factors they sought to remove. Every quality is at its most powerful when it is most concentrated; sweetness at its sweetest, bitterness at its bitterest, sharpness at its sharpest and so forth. The existence of such qualities in the body of man was perceived together with their harmful effects. There exists in man saltness, bitterness, sweetness, sharpness, astringency, flabbiness and countless other qualities having every kind of influence, number and strength. When these are properly

mixed and compounded with one another, they can neither be observed nor are they harmful. But when one is separated out and stands alone it becomes both apparent and harmful. Similarly, the foods which are unsuitable for us and harm us if eaten, all have some such characteristic; either they are bitter or salt or sharp or have some other strong and undiluted quality. For that reason we are disturbed by them, just as similar qualities when retained in the body harm us. Those things which form the ordinary and usual food of man, bread and barley-cakes and the like, are clearly farthest removed from those things which have a strong or strange taste. In this way they differ from those that are prepared and designed for pleasure and luxury. The simple foods least often give rise to bodily disturbance and a separation of the forces located there. In fact, strength, growth and nourishment come from nothing but what is well mixed and contains no strong nor undiluted element.

15. I am utterly at a loss to know how those who prefer these hypothetical arguments and reduce the science to a simple matter of 'postulates' ever cure anyone on the basis of their assumptions. I do not think that they have ever discovered anything that is purely 'hot' or 'cold', 'dry' or 'wet', without it sharing some other qualities. Rather, I fancy, the diets they prescribe are exactly the same as those we all employ, but they impute heat to one substance, cold to another, dryness to a third and wetness to a fourth. It would be useless to bid a sick man to 'take something hot'. He would immediately ask 'What?' Whereupon the doctor must either talk some technical gibberish or take refuge in some known solid substance. But suppose 'something hot' is also astringent, another is hot and soothing as well, while a third produces rumbling in the belly. There are many varied hot substances with many and varied effects which may be contrary one to another. Will it make any difference to take that which is hot and astringent rather than that which is hot and soothing, or even that which is cold and astringent or cold and soothing? To the best of my knowledge the opposite is the case; everything has its own specific effect. This is not only true of the human body but is

seen in the various substances used for working hides and wood and other things less sensitive than flesh and blood. It is not the heating effect of the application which is so important as its astringent or soothing qualities and so on, and this is true whether the substance be taken internally or applied as an ointment or plaster.

16. I think cold and heat are the weakest of the forces which operate in the body, and for these reasons. So long as cold and heat are present together they are harmless, for heat is tempered by cold and cold by heat. But when the two principles are separated from each other then they become harmful. However, when the body is chilled, warmth is spontaneously generated by the body itself so there is no need to take special measures, and this is true both in health and in disease. For instance, if a healthy man cools his body by taking a cold bath or by any other means, the more he cools himself the warmer he feels when he resumes his garments and comes into shelter again. This is only true, of course, so long as he does not wholly freeze. Again if he should warm himself thoroughly with a hot bath or at a fire, and then go into a cool place, it will seem to be much colder than formerly and he will shiver more. Should anyone cool himself with a fan on a very hot day, the heat seems ten times more suffocating when the fan is stopped than if its cooling properties had not been used at all. Let us consider now a more extreme example. If people get their feet, hands or head frozen by walking through snow or from exposure to cold, think of what they suffer from burning and irritation at night when they are wrapped up and come into a warm place; in some cases blisters come up like those formed by a burn. But these things do not happen before they get warm. This shows how readily each of this pair replaces the other. There are countless other examples I might give to illustrate this subject. Is it not true of sick men that those who have the severest chill develop the highest fever? And even when the fever abates its fury a little, the patient remains very hot. Then subsequently as it passes through the body, it finishes in the feet, that is the first part of the body to be attacked by the chill and the part which remained cold the longest. Again

when the patient sweats as the fever falls, he feels much colder than if he had not had the fever at all. What great or fearful effect, then, can a thing have when its opposite appears of itself with such speed and removes any effect that the former may have had? What need is there, also, for further assistance when nature neutralizes the effect of such an agent spontaneously?

17. Some may raise the objection that the fever of patients suffering from *causus*, pneumonia or other serious diseases does not rapidly decline. Neither in such cases is the fever intermittent. I think that such observations constitute a good proof of my own view that a high temperature is not the only element of a fever nor the only cause of the weak constitution of a febrile patient. May it not be said of a thing that it is both bitter and hot, sharp and hot, salt and hot, and countless other combinations both with heat and cold? In each combination, the effect of any two qualities acting together will be different. Such qualities may be harmful but there is as well the heat of physical exertion which increases as the strength increases and has no ill effects.

18. The truth of this may be demonstrated by the following consideration of certain signs. An obvious one, and one we have all experienced and shall continue to do so, is that of the common cold. When we have a running at the nose and there is a discharge from the nostrils, the mucus is more acrid than that which is present when we are well. It makes the nose swell and renders it hot and extremely inflamed, if you apply your hand to it. And if it lasts a long time, the part, being fleshless and hard, becomes ulcerated. The fever does not fall when the nose is running, but when the discharge becomes thicker, less acrid, milder and more of its ordinary consistency. Similar changes may be seen as the result of cold alone, but the same observations can be made. There is the same change from cold to hot and hot to cold and the changes take place readily and do not require any process of 'digestion'. I assert too that all other illnesses that are caused by acrid or undiluted humours within the body follow a similar course; they subside as these humours become less potent and are diluted.

19. Those humours which affect the eyes are very acrid and cause sores upon the eyelids; sometimes they cause destruction of the cheeks and the parts beneath the eyes. The discharge destroys anything it may touch, even eating away the membrane which surrounds the eye. Pain, heat and swelling obtain until such time as the discharges are 'digested' and become thicker and give rise to a serum. The process of 'digestion' is due to their being mixed and diluted with one another and warmed together. Again, the humours of the throat which cause hoarseness and sore throats, those of erysipelas or pneumonia are at first salt, moist and acrid and during this phase the maladies flourish. But when the discharges become thicker and milder and lose their acridity, the fevers cease as well as the other effects of the disease which are harmful to the body. The cause of these maladies is found in the presence of certain substances, which, when present, invariably produce such results. But when the nature of these substances becomes changed, the illness is at an end. Any abnormal condition which arose purely as a result of heat or cold and into which no other factor entered at all would be resolved when a change occurred from hot to cold or vice versa. However, the changes which take place really occur in the manner I described above. All the ills from which man suffers are due to the operation of 'forces'. For instance, if a sufferer from biliousness, complaining of nausea, fever and weakness, gets rid of a certain bitter material which we call yellow bile either by himself or with the assistance of purging, it is evident how he gets rid of both the fever and the pain at the same time. As long as this material is unabsorbed and undiluted, no device will terminate either the pain or the fever. When there are pungent rust-coloured acids present in the body, there is frenzy and severe pain in the bowels and in the chest and distress which cannot be cured until they have been purged of the acrid humours responsible and their poisonous effects neutralized by being mixed with other fluids. It is in the processes of digestion, change, dilution or thickening by which the nature of a humour is altered that the causes of disease lie. It is for this reason that the occurrence of crises and the periodicity of certain diseases are

so important. It is most improper that all these changes should be attributed to the effects of heat and cold, for such principles are not subject to degeneration or thickening. The changes of disease cannot be due to the effect of varying mixtures of such principles, for the only thing that will mix with heat and reduce its warmth is coldness and vice versa. The various forces in the body become milder and more health-giving when they are adjusted to one another. A man is healthiest when these factors are co-ordinated and no particular force predominates.

20. I think I have discussed this subject sufficiently, but there are some doctors and sophists who maintain that no one can understand the science of medicine unless he knows what man is; that anyone who proposes to treat men for their illnesses must first learn of such things. Their discourse then tends to philosophy, as may be seen in the writings of Empedocles and all the others who have ever written about Nature; they discuss the origins of man and of what he was created. It is my opinion that all which has been written by doctors or sophists on Nature has more to do with painting than medicine. I do not believe that any clear knowledge of Nature can be obtained from any source other than a study of medicine and then only through a thorough mastery of this science. It is my intention to discuss what man is and how he exists because it seems to me indispensable for a doctor to have made such studies and to be fully acquainted with Nature. He will then understand how the body functions with regard to what is eaten and drunk and what will be the effect of any given measure on any particular organ. It is not enough to say 'cheese is harmful because it produces pain if much of it is eaten'. One should know what sort of pain, why it is produced and which organ of the body is upset. There are many other harmful items of food and drink which affect the body in different ways. For example, the taking of large quantities of undiluted wine has a certain effect upon the body and it is recognized, by those who understand, that the wine is the cause and we know which organs are particularly affected. I want to show that the same sort of thing is true of other cases. Cheese, since that is the example I used, is not equally

harmful to all. Some can eat their fill of it without any unpleasant consequences and those whom it suits are wonderfully strengthened by it. On the other hand, there are some who have difficulty in digesting it. There must, then, be a difference in their constitutions and the difference lies in the fact that, in the latter case, they have something in the body which is inimical to cheese and this is aroused and disturbed by it. Those who have most of this humour and in whom it is at its strongest, naturally suffer most. If cheese were bad for the human constitution in general, it would affect everyone. Knowledge of this would avoid harm.

21. Both during convalescence as well as in the course of prolonged illnesses, complications are often seen. Some of them occur naturally in the course of the disease, others are occasioned by some chance happening. Most doctors, like laymen, tend to ascribe some such event to some particular activity that has been indulged in. In the same way they may ascribe something as being due to an alteration in their habits of bathing or walking or a change of diet, whether this is the actual case or not. As a result of jumping to conclusions, the truth may escape them. One must know with exactitude what is the effect of a bath or of fatigue indulged in at the wrong time. Neither such actions, nor eating too much, nor eating the wrong food will always produce the same effects; it depends upon other factors as well. No one who is unacquainted with the specific effects of such action on the body in different circumstances can know the results which follow and consequently he cannot make proper use of them as therapeutic measures.

22. I think it should also be known what illnesses are due to 'forces' and what to 'forms'. By 'forces' I mean those changes in the constitution of the humours which affect the working of the body; by 'forms' I mean the organs of the body. Some of the latter are hollow and show variations in diameter, being narrow at one end and wide at the other, some are elongated, some solid and round, some flat and suspended, some are stretched out, some large, some thick, some are porous and sponge-like. For instance, which type of hollow

organ should be the better able to attract and absorb moisture from the rest of the body: those which are all broad or those which are wide in part and narrow down? The latter kind. Such things have to be deduced from a consideration of what clearly happens outside the body. For instance, if you gape with your mouth wide open you cannot suck up any fluid, but if you pout and compress the lips and then insert a tube you can easily suck up as much as you like. Again, cupping glasses are made concave for the purpose of drawing and pulling the flesh up within them, and there are other examples of this kind of thing. Among the inner organs of the body, the bladder, the skull and the womb have such a shape and it is well known that these organs specially attract moisture from other parts of the body and are always filled with fluid. On the other hand organs which are more spread out, although they hold fluid which flows into them well, do not attract it to the same extent. Further, the solid and round organs neither attract it nor hold it because there is nowhere for the fluid to lodge. Those which are spongy and of loose texture such as the spleen, the lungs and the female breasts easily absorb fluid from the nearby parts of the body and when they do so become hard and swollen. Such organs do not absorb fluid and then discharge it day after day as would a hollow organ containing fluid, but when they have absorbed fluid and all the spaces and interstices are filled up, they become hard and tense instead of soft and pliant. They neither digest the fluid nor discharge it, and this is the natural result of their anatomical construction. The organs of the body that cause flatulence and colic, such as the stomach and chest, produce noise and rumbling. For any hollow organ that does not become full of fluid and remain so but instead undergoes changes and movement, must necessarily produce noises and the signs of movement. The organs which are soft and fleshy tend to become obstructed and then they are liable to sluggishness and fullness. Sometimes an organ which is diseased comes up against some flat tissue which is neither strong enough to resist the force of the swollen organ nor sufficiently mobile to accommodate the diseased organ by yielding. For instance, the liver is tender, full-blooded and solid and on account of these qualities is resistant to the movement of other organs. Thus wind, being obstructed by it, becomes more forceful and attacks the thing which obstructs it with greater power. In the case of an organ such as the liver, which is both full-blooded and tender, it cannot but experience pain. For this reason, pain in the hepatic area is both exceedingly severe and frequently encountered. Abscesses and tumours also occur very commonly here, as well as beneath the diaphragm. This latter condition, although less common, is more serious. The extent of the diaphragm is considerable and is opposed to other organs; nevertheless, its more sinewy and stronger nature makes it less liable to pain although both pains and tumours may occur in this region.

23. There are many individual variations in the shape of the different organs of the body from one person to another and they react differently both in health and in disease. There are large and small heads; thin and thick, long and short necks. The belly may be large and round; the chest narrow and flat. There are countless other differences and the effects of such variation must be known so that one can understand the exact cause when they become diseased. Only thus can proper care be given.

24. Again, the effect of each type of humour on the body must be learnt and, as I said before, their relationships with one another must be understood. I mean this sort of thing: if a sweet humour should change its nature, not by admixture with something else but spontaneously, what characteristic would it show? Bitter, salt, astringent or sharp? Sharp, I fancy. A sharp humour, compared with the others, would be specially inimical to the digestion of food. At least it would be so if, as we believe, a sweet humour is the most suited.

Thus, if anyone were able to light upon the truth by experiment outside the body, he would always be able to make the best pronouncements of all. The best advice is that which is least unsuitable.

EPIDEMICS, BOOK I

*The Hippocratic Corpus contains seven books of a physician's case notes.
They consist of descriptions of both individual cases and diseases epidemic in a
specified place in a given period. Books I and III appear to be the earliest
and most interesting, and date from the fifth century B.C.; the other five
books are probably the work of at least two later authors.*

(i)

1. There was much rain in Thasos about the time of the
autumnal equinox and during the season of the Pleiads. It fell
gently and continuously and the wind was from the south.
During the winter, the wind blew mostly from the south;
winds from the north were few and the weather was dry. On
the whole the winter was like springtime; but the spring was
cold with southerly* winds and there was little rain. The
summer was for the most part cloudy but there was no rain.
The etesian winds were few and light and blew at scattered
intervals.†

Although the climate was generally southerly and dry, in the
early spring there was a northerly spell, the very opposite of
the previous weather. During this time a few people contracted *causus* without being much upset by it, and a few had
haemorrhages but did not die of them. Many people suffered
from swellings near the ears, in some cases on one side only,
in others both sides were involved. Usually there was no fever
and the patient was not confined to bed. In a few cases there
was slight fever. In all cases the swellings subsided without
harm and none suppurated as do swellings caused by other
disorders. The swellings were soft, large and spread widely;
they were unaccompanied by inflammation or pain and they
disappeared leaving no trace. Boys, young men and male

* Possibly a copyist's error for 'northerly'.
† The etesian winds blow from the north-west for forty days in the
summer.

87

adults in the prime of life were chiefly affected and of these, those given to wrestling and gymnastics were specially liable. Few women took it. Many patients had dry, unproductive coughs and hoarse voices. Soon after the onset of the disease, but sometimes after an interval, one or both testicles became inflamed and painful. Some had fever, but not all. These cases were serious enough to warrant attention, but for the rest, there were no illnesses requiring care.

2. During the period beginning in early summer and lasting into the winter, many patients with long-standing consumption took to their beds, for in many cases in which the diagnosis had been dubious, it was then confirmed. Some whose constitution showed a tendency towards consumption first began to suffer from the disease at that time. Many died including most of the latter, and of those who took to their beds I doubt if any survived even a moderate time. Death occurred more quickly than is usual in such cases. Other diseases, even the longer ones and those accompanied by fever, proved neither serious nor fatal; these will be described later. Only consumption was widespread and caused a large number of deaths.

In the majority of cases the course of the disease was as follows. There was fever, attended by shivering, of continuous, severe and usually non-remittent type. However the fever showed some variation of tertian periodicity. Thus one day the fever would be less severe, the next day the fever would be higher and so on, but in general becoming worse with time. The patients showed continuous sweating but the whole of the body was not involved. The extremities were often cold and could be warmed only with difficulty. Their stomachs were disordered and the stools small, bilious, not homogeneous, fluid and pungent, causing the patient to get up frequently. The urine was either thin, colourless and undigested, or thick with a slight sediment which did not settle easily but was, as it were, raw and unripe. Cough was slight but frequent, and little was coughed up and that only with difficulty. In the most violent cases, there was no progress towards ripening of the sputum and the patients continued to cough it up raw. In most of these cases the throat was painful, red and inflamed from the first and continued so. The stools were small, thin and pungent. The patients rapidly became worse and wasted away, refusing to take food and having no thirst. Many became delirious shortly before death.

3. While it was still summer and during the autumn there were also many cases of fever apart from consumption. These were continuous but not violent and, though those affected were ill for a long time, they suffered nothing in other respects for their stomachs generally remained in good order and they took no harm worth speaking of. Usually the urine was clear and of a good colour but thin, becoming ripened later about the time of the crisis. There was not too much coughing, nor did the patients have much trouble with the cough. They retained their appetites and it was quite permissible to give them food. Generally they were only slightly ill and showed none of the fevers attended with shivering suffered by consumptive patients, and little sweating. The paroxysms of fever were irregular, being at different intervals in different cases. In the shortest illnesses, the crisis occurred at about the twentieth day, in most cases at about the fortieth and in a number at about the eightieth. In some cases the fever resolved at a time different from those given above without reaching a crisis. In most of these the fever returned after a short interval and the crisis was reached in one of the usual periods. In many cases the malady was so protracted that it lasted into the winter.

Of all the diseases described in this section, only consumption proved fatal. The course of the remainder was smooth and no deaths occurred from the other fevers.

(ii)

4. There was unseasonably wintry weather in Thasos early in the autumn, and rainstorms suddenly burst to the accompaniment of northerly and southerly winds. This happened during the season of the Pleiads until their setting. The winter was northerly and there was much rain, with frequent heavy showers, as well as snow. Usually there were bright intervals

as well, and the cold weather could not be regarded as unseasonable. However, immediately after the winter solstice when the west wind usually begins to blow, the great storms returned with gales from the north, and snow and rain fell continuously from a sky full of racing clouds. This continued without a break until the equinox. The spring was cold with northerly winds accompanied by cloudy skies and much rain. The summer was not too scorching for the etesian winds blew steadily, but heavy rain followed again soon after the rising of Arcturus.

5. The whole year then was wet, cold and northerly. The winter was healthy for the most part but early in the spring a good few, in fact most people, fell sick. Ophthalmia was the first disease to make its appearance, being accompanied by pain, moist discharge and without suppuration. Many people had small styes break out which gave them trouble. Most relapsed but were finally cured late in the year towards autumn. During the summer and the autumn there were cases of dysentery, tenesmus and diarrhoea. Further, there were cases of bilious diarrhoea in which the stools were copious, thin, raw and sometimes watery and painful to pass. There were also many cases of discharges accompanied by strangury and a painful, bilious, watery discharge containing particles and pus. There was no disease of the kidneys (in these cases). Cases were seen in which there was vomiting of phlegm, bile and undigested food. Sweating occurred and the patients became flaccid all over. Often there was no fever and the patients were not confined to bed, but in many other cases which will be described there was fever. Those who exhibited all the symptoms to be mentioned were consumptive and suffered pain. During the autumn and on into the winter there were cases of continued fever, in a few cases *causus*, diurnal and nocturnal fevers, roughly tertian and exact tertian fevers, quartans and fevers of no regular form. There were many cases of each of the fevers about to be described.

6. *Causus* was the least frequent of these fevers and those affected by it suffered the least. There was no bleeding, except in a very few cases and then only very slight; nor was there

any delirium and in other respects all went well. The crisis was regularly attained, usually on the seventeenth day including the days of intermission. I knew of no case of *causus* which was fatal or which was complicated by brain-fever.

The tertian fevers were more common than *causus* and more troublesome. In all cases of this fever four periods regularly elapsed from the time the malady was contracted and the final crisis was reached after seven paroxysms. None suffered from relapses.

The quartan fevers showed, in many cases, their quartan nature from the start. In not a few cases, however, they emerged as quartans only on the departure of other fevers and ailments. As is usual they were long protracted, perhaps even more than usual.

There were many cases of quotidian, nocturnal and irregular fever; they lasted a long time whether the patients were confined to bed or not. In most cases the fever lasted through the season of the Pleiads until the winter. Often the disease was accompanied by convulsions, especially in the case of children when the fever was, at first, slight. Convulsions also sometimes followed the fever. Although these maladies were protracted, they were not usually serious unless the patient was already likely to die from some other cause.

7. The worst, most protracted and most painful of all the diseases then occurring were the continued fevers. These showed no real intermissions although they did show paroxysms in the fashion of tertian fevers, one day remitting slightly and becoming worse the next. They began mildly but continually increased, each paroxysm carrying the disease a stage further. A slight remission would be followed by a worse paroxysm and the malady generally became worse on the critical days. Although all patients suffering from these various fevers showed shivering fits at irregular times, such fits were least frequent and most irregular in patients with these continued fevers. Again, the fevers generally were attended with many fits of sweating but in cases of continued fever they were infrequent and brought harm rather than relief. In continued fever, too, the extremities were chilled and could

only be warmed with difficulty, and insomnia was followed by coma. In the fevers generally, digestion was disturbed and difficult but this was most marked in these cases of continued fever. In them, too, the urine was either: (*a*) Thin, raw and colourless, becoming slightly more concocted at a crisis, (*b*) thick, but cloudy rather than forming sediment, or (*c*) of small quantity, bad and forming a raw sediment. Urine of this last variety was the most serious. Cough accompanied the fever, but I have no instance to record of a cough being either harmful or helpful.

8. These various symptoms [in cases of continued fever] were usually long-lasting, distressing and occurred without any order or regularity. In the majority of cases there was no crisis, whether or not the case was desperate. Some showed a brief respite but a relapse quickly followed. In a few cases a crisis occurred not earlier than the eightieth day but in some of these there was a relapse so that the majority of cases lasted on into the winter. Generally the disease resolved without a crisis, and this absence of crisis was equally marked both in those who recovered and those who did not.

These cases of [continued] fever, although they showed this characteristic of not reaching a crisis, were otherwise very varied. The most important and most ominous sign, which in the end was seen in most cases, was complete loss of appetite. This was specially marked in those whose condition was already desperate in other respects. Further, these febrile patients showed no greater desire for water than they would normally. Abscesses formed in those cases in which the illness was very prolonged and attended by much pain and loss of weight. These were so large, in some cases, as to be insufferable, while others were too small to be any good so that the patient soon returned to his former condition and deterioration was hastened.

9. This disease was commonly complicated by dysentery, tenesmus and diarrhoea. Some patients showed dropsy with or without these other symptoms. A violent attack of any of these complications quickly proved fatal; less severe attacks did no good. The disease was sometimes accompanied by

transient eruptions, quite out of keeping with the scale of the disease, and by swellings near the ears which slowly absorbed and signified nothing. In some cases similar swellings occurred in the joints, especially the hips. Usually in such cases a crisis was reached in a few days and the swelling disappeared, only to regain a hold and quickly return to the original state.

10. All the diseases described caused death, but the greater number was among those suffering from this continued fever and especially children, including infants, older children (eight and ten year olds) and those approaching puberty. The complications described latterly were invariably accompanied by the general symptoms of the disease as at first described. On the other hand, many who suffered from these symptoms did not suffer the complications. The most important sign, and the only good one, which saved many who were in the gravest danger, was when strangury occurred and a local abscess was produced. Strangury, in these conditions, occurred most commonly at the ages mentioned above but in many other cases as well, both when the patient was confined to bed and when he was not. In those who showed this symptom, a rapid and violent change took place, for the belly, even though it might have been malignantly moist, rapidly became firm, while the patient's appetite for all kinds of food fully returned and the fever thereafter was mild. But even in these cases, the strangury lasted long and was painful, the urine being copious, thick, varied, red, mixed with pus and passed with pain. Nevertheless, all these patients recovered; I do not know of one who died.

11. Whenever there is danger, watch out for all ripe discharges that flow from every part of the body at their due times and for favourable and critical abscess formation. Ripeness shows that the crisis is at hand and that recovery is certain. On the other hand, what is raw and immature, as well as unfavourable abscess formation, denotes the failure to reach a crisis, pain, prolongation of the malady, death or relapse. To decide which course is likely you must consider other things too. Consider what has gone before, recognize the signs before your eyes and then make your prognosis. Study

these principles. Practise two things in your dealings with disease: either help or do not harm the patient. There are three factors in the practice of medicine: the disease, the patient and the physician. The physician is the servant of the science, and the patient must do what he can to fight the disease with the assistance of the physician.

12. Headache and pains in the neck, and a feeling of heaviness accompanied by pain, may occur both in the presence and in the absence of fever. Those suffering from brain fever have convulsions and vomit brownish-red material and some of these die rapidly. If a patient has a pain in the neck, heaviness in the temples, dimness of vision and contraction of the hypochondrium without pain, he will have an epistaxis, both in *causus* and in other fevers. If his whole head is heavy or he feels heartburn and nausea, he will vomit bilious and phlegmatic matter. In these diseases, convulsions are more common among children, both convulsions and uterine pain occur in women, while older patients and all whose warmth is disappearing suffer from paralysis, madness or loss of sight.

(iii)

13. A little before the rising of Arcturus and during its season there were many violent rainstorms in Thasos accompanied by northerly winds. About the time of the equinox and until the setting of the Pleiads, the winds were southerly, so little rain fell. The winter was northerly with periods of drought, cold, high winds and snow. There were very severe storms at the time of the equinox. The spring was northerly, dry, with little rain, and it was cold. There was little rain at the time of the summer solstice but instead a severe cold spell set in and lasted till the [rising of the] Dog Star. Thence, until the [rising of] Arcturus, the summer was hot. This hot spell began suddenly and was both continuous and severe. There was no rain and the etesian winds blew. About the time of Arcturus, southerly rains began and continued until the equinox.

14. Under such circumstances, cases of paralysis started to

appear during the winter and became common, constituting an epidemic. Some cases were swiftly fatal. In other respects, health remained good. Cases of *causus* were encountered early in the spring and continued past the equinox towards summer. Most of those who fell sick in the spring or at the very beginning of summer recovered, though a few died. In the autumn, when the rains came, the disease was more fatal and the majority of those that took it died.

It was a peculiarity of *causus* that a good copious epistaxis often proved a cure, and I do not know of any in these circumstances who died if they had a good epistaxis. For Philiscus, Epameinon and Silenus had a small epistaxis on the fourth and fifth days; they died. Most of those who were sick had shivering attacks about the time of the crisis, especially those who did not have epistaxis. Such patients also had attacks of sweating.

15. Some cases of *causus* developed jaundice on the sixth day and these were assisted by the evacuation of urine, abdominal disturbance or by a profuse haemorrhage, such as Heracleides (who lay at Aristocydes' house) had. Moreover in this case he did not only have epistaxis but trouble in the belly and diuresis as well. He reached a crisis on the twentieth day. The servant of Phanagoras was not so lucky; he had none of these things happen to him and he died.

Most patients suffered from haemorrhage and especially was this the case in youths and young men. Indeed, of the latter who did not have a haemorrhage, most died. In older people the disease turned to jaundice or their bellies were upset, as was the case of Bion, who lay at Silenus' house. During the summer, dysentery became epidemic and those who had not recovered by that time had their sickness end up as a sort of dysentery, even when they had had a haemorrhage. This happened to Myllus and to Erato's slave whose illness, after a copious haemorrhage, turned to a sort of dysentery; they survived.

In fact, in this disease, this fluid was peculiarly abundant. Even those who did not bleed about the time of the crisis suffered pain and passed thin urine at this time and then began to bleed slightly about the twenty-fourth day, and there was pus mixed with the blood. In the case of Antiphon the son of Critobulus this finally ceased and the ultimate crisis was reached about the fortieth day. Such cases showed hard swellings near the ears which absorbed and were followed by a heaviness in the left flank and in the region of the iliac crest.

16. Many women were sick, but fewer women than men, and the disease in them was less fatal. Childbirth was often difficult and was followed by disease. These cases were specially fatal as, for instance, in that of the daughter of Telebulus, who died on the sixth day after giving birth. In most cases bleeding from the womb occurred during the fever and in many girls it occurred for the first time, but some had epistaxis. In some cases both bleeding from the womb and epistaxis were observed. For instance, the daughter of Daitharses who was a virgin not only had uterine bleeding for the first time then but also had a violent discharge of blood from the nose. I know of no case which proved fatal if either of these complications ensued. So far as I know, all who fell ill while pregnant aborted.

17. Generally in this disease the urine was of good colour but thin with a slight sediment. The belly was disordered, the stools being thin and bilious. In many cases, after a crisis had been reached for other disorders, the malady ended up as dysentery, as happened to Xenophanes and Critias. I will record the names of those patients who had watery, copious and fine urine, even after the crisis, with a healthy sediment, and who had a favourable crisis in other respects too. They were Bion, who lay at the home of Silenus; Cratis, who was at the house of Xenophanes; the slave of Areto, and the wife of Mnesistratus. All these subsequently suffered from dysentery.

About the time of Arcturus many reached the crisis on the eleventh day and they did not suffer the expected relapses. About this time, especially in children, the malady was associated with coma and these cases were the most rarely fatal of all.

18. *Causus* lasted on to the equinox, up to the setting of the Pleiads, and even into the winter. But at this time brain fever

became prevalent and most of its victims died. A few similar cases were also seen during the summer. Those suffering from fever of the *causus* type which proved fatal showed certain additional symptoms even at the beginning of the illness. High fever attended the beginning of the illness along with slight shivering fits, insomnia, thirst, nausea and a little sweating about the forehead and over the clavicles (in no case all over), much delirium, fears and despondency, while the extremities such as the toes were chilled, but especially the hands. Paroxysms occurred on even days. Generally, pain was greatest on the fourth day and the sweat was cold. Their extremities did not regain warmth but remained cold and livid, and they no longer suffered from thirst. They passed little urine, which was black and fine, and became constipated. In none of these cases was there a discharge of blood from the nose but only a few drops. Nor did these cases show any remission but died on the sixth day, sweating. Those patients who developed brain fever had all the above symptoms, but the crisis usually took place on the eleventh day. Where brain fever was not present at the beginning but appeared on the third or fourth day, the crisis did not take place until the twentieth day. In these the illness was moderate in its severity at first but became severe about the seventh day.

19. The disease was very widespread. Of those who contracted it death was most common among youths, young men, men in the prime of life, those with smooth skins, those of a pallid complexion, those with straight hair, those with black hair, those with black eyes, those who had been given to violent and loose living, those with thin voices, those with rough voices, those with lisps and the choleric. Many women also succumbed to this malady. During this epidemic there were four signs which betokened recovery: a considerable epistaxis, a copious discharge of urine that contained a lot of favourable sediment, biliousness and disorders of the belly coming on at a favourable time, or if there were dysentery. In many cases the crisis was not reached upon the appearance of one of the symptoms described, but instead the symptoms appeared successively and the patients seemed to be in a very

bad way. But in every such case they recovered. All these symptoms were seen in women and girls and if either any of them appeared or there was copious uterine haemorrhage, it proved their salvation and brought on the crisis. I do not know of any woman who died in which one of these signs had properly appeared. However, the daughter of Philo had a severe epistaxis, but she dined rather intemperately on the seventh day of her sickness; she died.

If a patient weeps in spite of himself in acute fever of the type of *causus*, you must expect an epistaxis, even if there is no other reason to expect a fatal outcome. If a patient be poorly, it portends not haemorrhage but death.

20. Swelling near the ears which sometimes accompanied fevers did not always subside or suppurate when the fever was resolved by crisis, but subsided following bilious diarrhoea or dysentery or by the formation of sediment in the urine as happened in the case of Hermippus of Clazomenae. The times of the crises in these fevers, which is the thing by which we distinguish them, were sometimes an even and sometimes an odd number of days. Thus, two brothers who lay near the summer residence of Epigenes fell sick at the same time. The elder reached a crisis on the sixth day, the younger on the seventh. Both relapsed at the same time following an intermission of five·days. After the relapse they reached a crisis together on the seventeenth day from the beginning of the illness. Generally the crisis was attained on the sixth day and, following an intermission of six days, a second crisis was reached on the fifth day of the relapse. In some cases the crisis took place on the seventh day, the intermission lasted seven days and the relapse reached its crisis in three days. In others, the crisis occurred on the seventh day, and a second on the seventh day of the relapse which followed three days' intermission of fever. In some cases a crisis took place on the sixth day, the remission lasted six days and this was followed by three days' relapse, a remission of one day, a relapse of one day and finally the crisis. This happened to Evagon, the son of Daitharses. In other cases, a crisis took place on the sixth day, the remission lasted seven days with crisis on the fourth day

of the relapse, as happened to the daughter of Aglaïdas. The majority of those who caught this epidemic passed their illness in the manner described and I know of none that survived who did not have a relapse in the normal way. All who had relapses of this sort recovered so far as I know. Further, to my knowledge, none whose malady proceeded in this manner subsequently suffered a return of the disease.

21. In these fevers, death usually took place on the sixth day, as happened in the case of Epaminondas, Silenus and Philiscus the son of Antagoras. Those who had a swelling near the ears had a crisis on the twentieth day but in all cases it subsided without suppuration, being voided in the urine. Cratistonax, who lived near the temple of Heracles, and the servant girl of Scymnus the fuller developed abscesses; they died. In some cases the crisis was on the seventh day, and following nine days' remission the fever recurred and reached its crisis on the fourth day after the recurrence; this happened in the case of Pantacles who lived near the temple of Dionysus. Sometimes the first crisis was on the seventh day, the remission lasted six days and a crisis was reached on the seventh day of the recrudescence; this happened in the case of Phanocritus who lay at the home of Gnathon the fuller.

22. During the winter, from about the time of the winter solstice till the equinox, *causus* and brain fever continued, and there were many deaths. There was, however, a change in the periods at which the crisis occurred, it taking place usually on the fifth day from the beginning of the illness. A remission of four days would be followed by a relapse with the crisis on the fifth day, that is on the fourteenth day of the illness. Most of those who behaved in this way were children, but it happened occasionally in adults. In some cases a crisis occurred on the eleventh day, a relapse on the fourteenth and the final crisis on the twentieth. But if shivering fits supervened about the twentieth day, the crisis took place on the fortieth. Most patients suffered from shivering fits about the time of the first crisis, and those who had them then also had them at the time of the crisis of the relapse. Very few had shivering fits during the spring, more had them during the summer, still more

during the autumn but by far the greatest number during the winter. Cases of haemorrhage gradually ceased.

23. The factors which enable us to distinguish between diseases are as follows: First we must consider the nature of man in general and of each individual and the characteristics of each disease. Then we must consider the patient, what food is given to him and who gives it – for this may make it easier for him to take or more difficult – the conditions of climate and locality both in general and in particular, the patient's customs, mode of life, pursuits and age. Then we must consider his speech, his mannerisms, his silences, his thoughts, his habits of sleep or wakefulness and his dreams, their nature and time. Next, we must note whether he plucks his hair, scratches or weeps. We must observe his paroxysms, his stools, urine, sputum and vomit. We look for any change in the state of the malady, how often such changes occur and their nature, and the particular changes which induce death or a crisis. Observe, too, sweating, shivering, chill, cough, sneezing, hiccough, the kind of breathing, belching, wind, whether silent or noisy, haemorrhages and haemorrhoids. We must determine the significance of all these signs.

24. Some fevers are continuous, others come at day and remit at night; others for the night, remitting by day. There are sub-tertian, tertian and quartan fevers, five-day, seven-day and nine-day fevers. The most severe, serious, troublesome and fatal maladies produce continued fevers. The safest, easiest to bear and yet longest of all is the quartan fever, not only from its own nature but also because it puts an end to other serious illnesses. What is termed sub-tertian fever can occur in acute illnesses, and it is the most fatal of all; but consumption and other protracted diseases are especially prone to take this form. Nocturnal fever is not especially fatal but it is long drawn out. Diurnal fever is longer still and sometimes leads to consumption. The seven-day fever is long-lasting but not fatal. The nine-day fever is still longer but not fatal. An exact tertian fever soon produces a crisis and is not fatal. The five-day fever is the worst of all, for when it comes on before

consumption or when the patient be already consumptive, it is fatal.

25. Each of these fevers has its characteristics, both in the nature of the fever and the spacing of the paroxysms. For example, a continued fever in some cases rapidly attains its height and then the fever diminishes as the crisis is approached and is passed. In other cases it begins gently without producing obvious signs, increasing day by day paroxysmally until at the crisis it fairly shines out. In other fevers, the start is mild but the fever increases in paroxysms to its height and then persists until the crisis be reached and passed. These different signs may be displayed by any fever or sickness. Before deciding on treatment, you must also consider the patient's mode of life. There are also many other signs of importance to be considered in these conditions; some have already been described elsewhere, others await description. They must all be taken into account in deciding whether the patient will have a short or protracted illness, a fatal outcome or cure. Similarly these things will have to be considered in deciding what treatment to adopt and the nature, quantity and time of administration of medicaments.

26. Fevers attended by paroxysms at even numbers of days, reach their crisis also in an even number; if the paroxysms are on odd days, so is the crisis. The first period [of fever] in those maladies which reach the crisis in an even number of days is 4, 6, 8, 10, 14, 20, 24, 30, 40, 60, 80, or 120 days. If the crisis be reached in an odd number, then the first period lasts 3, 5, 7, 9, 11, 17, 21, 27 or 31 days. It must be noted that if a crisis occurs on any other day than those mentioned, there will be a relapse and also it may prove a fatal sign. One must pay attention to these days which have been specified in the course of a particular fever and realize that on them a crisis may take place leading to recovery or death, to improvement or to deterioration. In irregular fevers, quartans, five-, seven-, and nine-day fevers, one must also take note of the periodicity with which the crises occur.

FOURTEEN CASES

(i)

Philiscus lived near the city wall. He took to his bed on the first day of his illness with high fever and sweating and passed an uneasy night.

On the second day all the symptoms became more pronounced, and later in the day his bowels were well opened following the administration of an enema. He spent a quiet night.

Third day: in the early morning and until midday he appeared to be without fever: towards evening, a high fever with sweating, thirst, a parched tongue and he passed dark urine. Spent a restless night without sleeping and was quite out of his mind.

Fourth day: symptoms more pronounced; urine dark. An easier night; urine a better colour.

Fifth day: about midday a slight epistaxis of pure blood; urine not homogeneous but containing globular particles suspended in it, like semen, which did not settle. Following the giving of a suppository, passed small stools with flatulence. Night uneasy, short snatches of sleep, talking, delirium, extremities all cold and could not be warmed, passed dark urine, slept a little towards daybreak, lost his voice, cold sweating, extremities livid.

About midday on the sixth day he died.

Throughout he took deep infrequent breaths as if consciously controlling his breathing. The spleen was enlarged and presented as a round lump; cold sweats all through the illness. The paroxysms were on even days.

(ii)

Silenus lived on the flat ground near Evalcidas' place. He took a fever as the result of fatigue, drink and untimely exercise. He started with a pain in the loins, heaviness of the head and retraction of the neck.

On the first day, stools copious, bilious and not homogeneous; frothy and dark-coloured. Urine dark with a dark sediment; thirst, tongue dry; did not sleep that night.

Second day: high fever, stools more copious, thinner and frothy; urine dark. Passed a restless night, slight delirium.

Third day: all symptoms more pronounced; contraction of the hypochondrium on both sides extending as far as the navel, somewhat flabby underneath the contraction. Stools thin and somewhat dark in colour; urine cloudy and rather dark. No sleep that night, much talking, laughter, singing, could not be restrained.

Fourth day: condition unchanged.

Fifth day: stools unmixed, bilious, smooth and fatty; urine thin and clear; showed slight signs of understanding.

Sixth day: slight sweating about the head, extremities cold and livid, much tossing about, complicated with constipation and suppression of urine, high fever.

Seventh day: lost his voice, extremities could no longer be warmed, anuria and retention of stools continued.

Eighth day: cold sweating all over, accompanied by spots. These were red, round and small like those of acne which did not go down. A thin copious stool, as if undigested, was passed with difficulty following a small enema. The urine was passed with pain and was pungent; extremities became slightly warm, periods of light sleep, signs of coma, loss of voice, urine thin and clear.

Ninth day: condition unchanged.

Tenth day: would not take drink, comatose, periods of light sleep; stools the same, passed a large quantity of rather thick urine which formed a white sediment like barley-meal on standing. Extremities again cold.

Eleventh day: he died.

From the beginning and throughout the illness he took deep infrequent breaths. Continuous pulsation of the hypochondrium. Age about twenty.

(iii)

Herophon suffered from a high fever. Stools small with tenesmus at first; afterwards he passed thin bilious matter rather frequently. He could not sleep, urine dark and thin.

Early on the fifth day, became deaf, all the symptoms were more pronounced, the spleen became enlarged, the hypochondrium contracted; he passed a small quantity of dark matter from his bowels. He was delirious.

Sixth day: babbling at random, at night sweating, became cold, remained delirious.

Seventh day: became cold, thirsty, out of his mind. Regained control of his mind during the night and slept.

Eighth day: fever; the spleen was reduced in size and he was wholly lucid. He felt a pain at first in the groin on the same side as the spleen, later on pains in the calves of both legs. Passed a comfortable night. Urine of better colour with a slight sediment.

Ninth day: sweating, the crisis was reached and the fever left him.

On the fifth day after this a relapse occurred. The spleen immediately enlarged; high fever, deafness again. On the third day of the relapse the spleen became reduced, the deafness less; pain in the legs; sweating during the night. A crisis was reached about the seventeenth day. There was no delirium during the relapse.

(iv)

In Thasos, the wife of Philinus, having given birth to a daughter fourteen days previously, the lochia being normal and the patient doing well, was taken ill with a fever accompanied by rigors. At first she felt a pain in the heart and in the right hypochondrium; pains in the genitalia; the lochia ceased. When a pessary was applied these pains were eased, but the pains in the head, neck and loins remained. No sleep, extremities cold, thirst, belly dried up, passed small stools, urine thin and of bad colour at first.

On the sixth night, a long attack of delirium followed by lucidity.

Seventh day: thirst, small bilious dark-coloured stools.

Eighth day: further rigors, high fever, a large number of painful convulsions, much talking at random. An enema produced copious bilious stools. Sleep not possible.

Ninth day: convulsions.

Tenth day: slight recovery of senses.

Eleventh day: slept, remembered everything, but very soon became delirious again. Passed much urine spontaneously accompanied by frequent convulsions; the urine was thick and white, looking like urine with a sediment which has been stirred up, but when it was left standing a long time it did not in fact produce a sediment and resembled in colour and thickness the urine of cattle. I myself examined the urine.

About the fourteenth day, a throbbing throughout the body, a lot of talking, slight lucidity rapidly followed by renewed delirium.

About the seventeenth day she became speechless and died on the twentieth day.

(v)

The wife of Epicrates, who lived near the statue of the founder, was said to have had a violent shivering fit about the time of childbirth, and could not get warm. The severe symptoms continued on the next day.

On the third day she gave birth to a daughter and parturition was normal.

On the second day after the delivery she had a high fever with pains in the heart and in the genitals which were eased by the application of a pessary. But she continued to suffer from pains in the head, neck and loins. Sleep was impossible. Her stools were small, thin, bilious and not homogeneous; urine thin and rather dark.

On the night of the sixth day of the fever she became delirious.

The symptoms became more pronounced on the seventh

day; insomnia, delirium, thirst, dark-coloured bilious stools.

Eighth day: rigors occurred, she slept rather more.

Ninth day: no change.

Tenth day: painful aching in the legs, pain recurred in the heart, headache, no delirium, slept more, bowels constipated.

Eleventh day: passed a lot of urine of good colour which produced a sediment. Was rather better.

Fourteenth day: rigors, high fever.

Fifteenth day: vomited yellow bilious matter rather frequently, sweated and became feverish; at night high fever again, urine thick with a white sediment.

Sixteenth day: a paroxysm, restless night, no sleep, delirium.

Eighteenth day: thirst, tongue parched, no sleep, a lot of delirium, pains in the legs.

About the twentieth day, slight rigors early in the morning, coma, slept restfully, vomited a small quantity of bilious dark matter, deafness at night.

About the twenty-first day, a painful heaviness all down the left side; coughed up a small amount. Urine thick, cloudy and reddish; did not form a sediment when left standing. An improvement in other respects but the fever continued. From the beginning of the illness the throat was painful and inflamed with the uvula retracted, and there was a pungent acrid salty discharge.

About the twenty-seventh day, no fever, a sediment in the urine, slight ache in the side.

About the thirty-first day the fever returned and the bowels were disordered, the stools bilious.

On the fortieth day she vomited a small quantity of bilious matter.

The final crisis and the end of the fever was reached on the eightieth day.

(vi)

Cleanactides, who lived on the hill above the temple of Heracles, was taken ill with an irregular fever. From the beginning he suffered from headache and pain in the left side; the rest of the body ached as it might from fatigue. The paroxysms of fever occurred in no regular sequence, but sometimes in one fashion, sometimes another. Sometimes there was sweating; sometimes not. Generally paroxysms were specially in evidence on the critical days.

About the twenty-fourth day he suffered from pain in the finger-tips and vomited, at first yellow bilious material, later rust-coloured matter. He was relieved of everything.

About the thirtieth day he began to bleed from both nostrils and slight epistaxis continued until the crisis. He did not suffer at all from lack of appetite, thirst or insomnia. Urine was thin, but not of a bad colour.

About the fortieth day he passed reddish urine with a large amount of red sediment. His condition improved. Subsequently the nature of the urine was varied; sometimes it had a sediment, sometimes none.

On the sixtieth day the urine contained a lot of white smooth sediment. All the symptoms decreased and the fever left him; the urine again became thin but of good colour.

On the seventieth day, fever followed by ten days' remission.

On the eightieth day, rigors and high fever; he sweated a lot, and the urine contained a red smooth sediment. Eventually the final crisis was reached.

(vii)

Meton suffered from a fever with a painful heaviness in the loins.

Second day: he took frequent drinks of water and his bowels were well opened.

Third day: heaviness in the head; the stools thin, bilious and reddish.

Fourth day: symptoms more pronounced; on two occasions he had an epistaxis from the right nostril. Passed a restless night; stools as on the third day; urine rather dark in colour and containing suspended particles which did not settle on standing.

Fifth day: a copious epistaxis from the left nostril of pure blood, sweating; the crisis was reached.

After the crisis, he had insomnia and delirium while his urine was thin and rather dark. He had his head bathed, slept and regained his wits. He did not relapse but he had many attacks of epistaxis following the crisis.

(viii)

Erasinus lived near the gully of Boötes. He was taken ill with fever after dinner and passed a disturbed night.

The first day was restful; was distressed during the night.

Second day: all symptoms more pronounced, delirium at night.

Third day: painful, much delirium.

Fourth day: worst of all so far; did not sleep at all at night. Visual hallucinations, delirium. These were followed by even more marked disturbances, feelings of fear and his illness was very severe.

Fifth day: in the early morning he became lucid and quite regained possession of his wits. But some time before noon he became mad and could not be restrained; extremities cold and somewhat livid. Suppression of urine. He died about sunset.

He had fever throughout the illness accompanied by sweating. The hypochondrium was distended and contracted only with pain. The urine was dark containing suspended globular particles which did not form a sediment on standing. His bowels remained open and he passed solid stools. Thirst throughout was not excessive. He had many convulsions accompanied by sweating about the time of death.

(ix)

Crito in Thasos had a violent pain in the foot which came on while walking; it started from the big toe. The same day he took to his bed with shivering, nausea and slight fever; at night he became delirious.

Second day: the whole foot became swollen; it was reddish about the ankle where there was some contraction and small black blisters appeared. He developed high fever and madness. He passed rather frequent unmixed bilious stools.

He died on the second day from the beginning of his illness.

(x)

The man from Clazomenae, who was dwelling near the well of Phrynichides, took a fever. From the beginning he had headache and aching pains in his neck and loins. Deafness was present from the start; sleep was impossible. High fever developed, the hypochondrium was distended but not excessively so, distension, tongue dry.

On the night of the fourth day he became delirious.

Fifth day: bad.

Sixth day: all symptoms more pronounced.

About the eleventh day a slight relief.

From the beginning of the illness until the fourteenth day the stools were thin, bile-stained and copious. Then the bowels became constipated. The urine throughout, although thin, was copious and of a good colour and contained scattered particles which did not settle.

About the sixteenth day he passed slightly thicker urine which formed a small amount of sediment. He became a little better and more lucid.

On the seventeenth day the urine became thin again and painful swellings by both ears appeared. Sleep was impossible; he was delirious and suffered from pain in the legs.

On the twentieth day he attained the crisis and lost his fever; he did not sweat, became completely lucid.

About the twenty-seventh day he had a violent pain in the right hip which quickly ceased. The swellings near the ears neither suppurated nor resolved, but remained painful.

About the thirty-first day, diarrhoea started; the stools were very watery with some evidence of dysentery. He passed thick urine. The swellings by the ears went down.

On the fortieth day his right eye ached and his sight became impaired; this trouble passed off.

(xi)

The wife of Dromeades, having given birth to a daughter and progressing in all other respects normally, was seized by rigors on the second day, accompanied with high fever.

A pain started on the subsequent day in the hypochondrium; nausea and shivering supervened. She did not sleep on succeeding days and was distraught. Breathing deep and slow, each breath immediately drawn back again.

On the second day after the rigors, her stools were normal; the urine thick, white and cloudy, of the appearance of urine with sediment which has been stirred up after standing a long while. But in her case no sediment was formed. She did not sleep at night.

Third day: about noon rigors, a high fever, urine as before, pain in the hypochondrium, nausea. A restless night with insomnia. Generalized cold sweats, but soon followed by warmth again.

Fourth day: slight relief of the pain in the hypochondrium, a heavy headache. Fell into a stupor; a slight epistaxis occurred. Tongue dry, thirst. Urine little, thin and oily. Slept a little.

Fifth day: thirst, nausea; character of urine unchanged, constipated. About noon much delirium followed quickly by a lucid phase. On going to stool she became comatose and chilled, slept during the night but was delirious.

Early on the sixth day she suffered from rigors followed quickly by fever and generalized sweating; the extremities were cold and she was delirious with a slow rate of breathing. After a little while convulsions supervened starting in the head and death soon followed.

(xii)

A man dined when heated and drank too much. During the night he vomited everything up; high fever, pain in the right hypochondrium with a soft inflammation therein; a restless night. Urine at first was thick and red and did not sediment when left to stand. Tongue dry but no excessive thirst.

Fourth day: high fever; generalized pain.

Fifth day: passed a large amount of smooth oily urine. High fever.

Sixth day: towards evening, much delirium. Insomnia.

Seventh day: all symptoms more pronounced. Urine as before; very talkative and could not be restrained. Following an enema he passed a liquid disturbed stool containing worms. A bad night; rigors in the early morning. High fever with warm sweating, then appeared to lose the fever. He did not sleep much and was chilled after sleeping. Expectoration. In the evening much delirium and after a short while vomited a small amount of black bilious matter.

Ninth day: chilled, much delirious babbling, insomnia.

Tenth day: pains in the legs, all symptoms more pronounced, delirium.

Eleventh day: death.

(xiii)

A woman who lived on the sea-front was seized with a fever while in the third month of pregnancy. She was immediately seized with pains in the loins.

On the third day, pain in the head, neck and round about the right clavicle. Very shortly the tongue became unable to articulate and the right arm was paralysed following a convulsion as happens in hemiplegia. Her speech was delirious. A restless night with insomnia; the bowels were disordered and the stools were small, bilious, and unmixed.

Fourth day: speech was indistinct but she was no longer paralysed, convulsions. Pains continued as before and there was a painful swelling near the hypochondrium. She did not sleep and was completely delirious. Bowels disordered; urine thin and not of a good colour.

Fifth day: high fever; pain in the hypochondrium, completely delirious. Stools bilious. At night she sweated and the fever left her.

Sixth day: lucid; a general improvement; pain persisted in the region of the left clavicle. Thirst, thin urine, insomnia.

Seventh day: trembling, fell into a stupor with slight delirium. Aching persisted about the clavicle and in the left upper arm, but in other respects her condition improved and she was fully lucid. The fever intermitted for three days.

On the eleventh day a relapse occurred with rigors and fever.

About the fourteenth day she vomited yellow bilious material rather frequently; sweating. Reached a crisis and the fever left her.

(xiv)

Melidia, who lay near the temple of Hera, began to suffer from violent headache and pain in the neck and chest. She was at once seized with a fever. There was some vaginal discharge. The pain from which she suffered was continuous.

Sixth day: fell into a coma, nausea, shivering, a rash appeared on the face, slightly delirious.

Seventh day: sweating. The fever declined but the pain remained. The fever returned; short snatches of sleep.

Urine was of good colour throughout but thin. The stools were thin, bilious and pungent, small in quantity, dark and offensive. The urine contained a white smooth sediment. Sweating took place. The final crisis was reached on the eleventh day.

EPIDEMICS, BOOK III

1 (i)

Pythion who lived near the temple of Earth suffered from twitching which began in the hands.

First day: high fever, delirium.

Second day: all symptoms more pronounced.

Third day: condition unchanged.

Fourth day: passed small, undigested, bilious stools.

Fifth day: all symptoms more pronounced, periods of light sleep, bowels constipated.

Sixth day: sputum not homogeneous and tinged red.

Seventh day: mouth distorted.

Eighth day: all symptoms more pronounced, fits of twitching continued. From the beginning of the illness until the eighth day, the urine was thin and pale, with cloudy matter in it.

Tenth day: sweated, sputum rather ripe, the crisis reached; urine rather thin about the time of the crisis. After the crisis, in fact forty days later, a peri-anal abscess formed which produced the symptom of strangury.

(ii)

Hermocrates, who lay near the new wall, took a fever. It started with a headache and pain in the loins. The hypochondrium was flabby and distended and the tongue was parched. He immediately became deaf, was unable to sleep, was thirsty but not excessively so, while the urine was thick and red and formed no sediment. There was some inflammatory matter in the stools.

Fifth day: passed thin urine which had particles suspended in it but which did not form a sediment; became delirious towards nightfall.

Sixth day: signs of jaundice; all symptoms more pronounced, not mentally lucid.

Seventh day: condition uneasy, urine thin as before. Condition remained more or less unchanged on the following days. About the eleventh day there was the appearance of general improvement but then coma supervened; he passed thicker reddish urine which was clear below. Slowly became lucid.

Fourteenth day: no fever, no sweating, slept, fully lucid. Urine much as before. A relapse with fever around about the seventeenth day. This was followed by high fever on succeeding days; delirium, thin urine. A second crisis occurred on the twentieth day; lost his fever, no sweating. The whole time he suffered from loss of appetite. He was fully lucid, but was unable to converse. Tongue dry but no thirst. Slept a little, then comatose. About the twenty-fourth day there was a further rise of temperature with diarrhoea. On the following days high fever continued and his tongue was parched.

Twenty-seventh day: died.

The patient's deafness lasted throughout his illness; the urine was thick and red forming no sediment, or else thin and colourless with suspended particles in it. The patient also lost his sense of taste.

(iii)

A man who dwelt in the park of Delearces suffered for a long time from heaviness of the head and right temporal pain. For some reason he took a fever and went to bed.

Second day: a small flow of pure blood from the left nostril. Bowels well opened; urine thin and not homogeneous, containing small suspended particles like barley-meal or semen.

Third day: a high fever; stools dark, thin and frothy with a livid sediment. Patient became stuporous; going to stool caused discomfort. Urine had a livid, somewhat sticky sediment.

Fourth day: vomited small quantities of yellow bilious matter and, after a while, a small quantity of rust-coloured material. There was a small haemorrhage of pure blood from the left side of the nose, stools and urine as before; sweating about the head and shoulders; spleen enlarged; pain in the region of the thigh; a rather flabby distension of the right hypochondrium; did not sleep at night; slight delirium.

Fifth day: stools larger, dark and frothy with a dark sediment; no sleep that night, delirium.

Sixth day: stools dark, oily, sticky and foul-smelling. Slept and was rather more lucid.

Seventh day: tongue dry, thirsty, did not sleep, delirious; urine thin, but not of a good colour.

Eighth day: stools small and dark, formed; slept and became lucid; thirsty, but not excessively so.

Ninth day: rigors supervened, a high fever, sweating, chilling, delirium, squint in the right eye; tongue dry, thirst, insomnia.

Tenth day: condition unchanged.

Eleventh day: became fully lucid, no fever, slept; urine was thin about the time of the crisis. He remained without fever for a period of two days but a relapse occurred on the fourteenth day, when he immediately became completely delirious and was sleepless all night.

Fifteenth day: urine muddy, resembling the appearance of urine which contains sediment after it is stirred up; a high fever, completely delirious, pain in the knees and calves. Passed dark stools after the application of a suppository.

Sixteenth day: urine thin with cloudy matter suspended in it; delirium.

Seventeenth day: in the morning the extremities were cold; he was wrapped up, had high fever, sweating all over; condition improved, became more lucid but not without fever; thirsty, vomited small quantities of yellow bilious matter; passed stools which after a little became dark, small in quantity and thin. Urine was thin but not of good colour.

Eighteenth day: was not lucid; comatose.

Nineteenth day: condition unchanged.

Twentieth day: slept, was fully lucid, sweating, no fever; not thirsty, urine thin.

Twenty-first day: slight delirium, somewhat thirsty, pain in the hypochondrium associated with palpitation in the umbilical region.

Twenty-fourth day: sediment in the urine; was fully lucid.

Twenty-seventh day: pain in the right hip, but in other respects doing very well; sediment in the urine. About the twenty-ninth day, pain in the right eye; urine thin. On the fortieth day, passed frequent white stools containing phlegmatic matter; sweated much all over and reached the final crisis.

(iv)

At Thasos, Philistes had a headache for a long time and, on falling into a state of stupor one day, took to his bed. As a result of drinking, continuous fever occurred and the pain became worse. It was at night that he first became hot.

First day: vomited small quantities of yellow bilious matter at first, afterwards more which was rust-coloured. Bowels were opened. An uneasy night.

Second day: deafness, high fever; the right hypochondrium was contracted and indrawn; urine thin, transparent and having a small amount of suspended particles in it resembling semen. He became mad about midday.

Third day: uneasy.

Fourth day: convulsions, a fit.

Fifth day: died in the morning.

(v)

Chaerion who lay at the house of Delias took a fever as the result of drinking. At once his head began to feel heavy and to ache; he had no sleep, bowels disordered with thin, rather bilious stools.

Third day: high fever, twitching of the head, particularly of the lower lip. After a while, rigor, convulsions, complete delirium; passed an uneasy night.

Fourth day: quiet, slept a little, delirious.

Fifth day: condition bad, all symptoms more pronounced, random babbling, an uneasy sleepless night.

Sixth day: condition unchanged.

Seventh day: rigors, high fever, sweating all over; the crisis reached. Throughout his stools were bilious, small in quantity and undigested. Urine thin, of poor colour and with a cloudy substance suspended in it. About the eighth day he passed urine of a better colour which had a small amount of white sediment; he became lucid and lost his fever and there was an intermission. On the ninth day a relapse occurred.

Fourteenth day: high fever.

Sixteenth day: vomited bilious yellow matter rather frequently.

Seventeenth day: rigors, high fever, sweating; reached a crisis and the fever left him. After the relapse and then the crisis, the urine was of good colour with a sediment. Delirium was absent during the relapse.

Eighteenth day: slight rise in temperature, slight thirst, urine thin with a cloudy substance suspended in it; slight delirium.

Nineteenth day: no fever, pain in the neck, sediment in the urine. The final crisis was reached on the twentieth day.

(vi)

The unmarried daughter of Euryanax took a fever. She suffered from no thirst throughout and did not take her food. Passed small stools; the urine thin, of small quantity and not of good colour. At the beginning of the fever she had pain around the anus. On the sixth day she did not sweat, as she was without fever, and reached a crisis. However, there was some slight suppuration about the anus, the abscess bursting about the time of the crisis. On the seventh day after the crisis, rigors occurred and there was slight fever with sweating. Subsequently she was always cold about the extremities. About the tenth day after the sweating occurred, she became delirious, but quickly recovered her lucidity again. They said it was through eating grapes. After an intermission of twelve days, she again became quite delirious, the bowels were disordered, the stools being bilious, unmixed, small in quantity, thin and pungent. The stools were passed frequently. She died on the seventh day following the last attack of her illness and a rash was present throughout, while the uvula was retracted. Fluxes were present, small and acrid. Although she had a cough it was unproductive. She had no appetite the whole time, nor did she wish for anything. She had no thirst and drank nothing worth mentioning. She was silent and would not talk. She was depressed and despaired of herself. There was also some sign of an inherited tendency to consumption.

(vii)

The woman who suffered from sore throat, who lived near Aristion's place, started first with her voice becoming indistinct. Her tongue was red and parched.

First day: shivering; high fever.

Third day: rigor, high fever; a hard reddish swelling on either side of the neck down to the chest, extremities cold and livid, respiration superficial. What she drank was regurgitated through the nostrils and she was unable to swallow. Stools and urine suppressed.

Fourth day: all symptoms more pronounced.

Fifth day: died.

(viii)

The lad who lay by the Liars' Market took a fever as the result of exhaustion, having exerted himself by running more than he was accustomed.

First day: bowels disordered with copious thin bilious stools; urine thin, rather dark; insomnia, thirst.

Second day: all symptoms more pronounced; stools more copious and unhealthy. No sleep; his mind was disordered; slight sweating.

Third day: uneasy, thirst, nausea, much tossing about, distress, delirium, extremities cold and livid; a somewhat flabby bilateral distension of the hypochondrium.

Fourth day: no sleep; condition deteriorated.

Seventh day: died.

Age about twenty.

(ix)

The woman at the house of Tisamenus was taken to her
bed feeling very ill with symptoms suggesting an attack of
ileus. Much vomiting; she could keep neither food nor drink
down. Pain in the hypochondrium; also pain lower down in
the belly proper. Constant colic. No thirst. Became warm,
but the extremities remained cold throughout; nausea; insomnia. Urine small in quantity and thin. Stools raw, thin and
small. It was impossible to do anything to help her; she died.

(x)

A woman of the household of Pantimides took a fever the
first day after a miscarriage. Tongue was parched; thirst,
nausea and insomnia, bowels disordered, the stools being thin,
copious and raw.

Second day: rigors, high fever, much purgation; did not
sleep.

Third day: pains more intense.

Fourth day: became delirious.

Seventh day: died.

The bowels were relaxed throughout, the stools being
watery, thin, raw and voluminous; urine little and thin.

(xi)

Another case of miscarriage about the fifth month resulted
in Hicetas' wife taking a fever. To begin with she was comatose
but later became wakeful and suffered from pain in the loins
and heaviness of the head.

Second day: bowels disordered with small, thin stools, at
first unmixed.

Third day: worse; did not sleep at night.

Fourth day: became delirious and suffered from fears and
from depression. Squint in the right eye; a small amount of
cold sweating about the head. Extremities cold.

Fifth day: all symptoms more pronounced; much delirious

talking, but she soon became lucid again. No thirst, insomnia; the stools were large in quantity, and unfavourable throughout; urine little in quantity, thin and rather dark. Extremities cold and somewhat livid.

Sixth day: no change.

Seventh day: death.

(xii)

A woman who lay near the Liars' Market, having given birth to a first-born male child after a difficult labour, took a fever. To start with she suffered from thirst, nausea and a slight ache in the heart; her tongue was parched and the bowels were disordered, her stools being thin and small. She did not sleep.

Second day: slight rigors, a high fever, a small amount of cold sweating about the head.

Third day: distressed; passed a large quantity of raw thin stools.

Fourth day: rigors, all symptoms more pronounced. Insomnia.

Fifth day: distressed.

Sixth day: no change; passed a large quantity of liquid stools.

Seventh day: rigors, high fever, thirst, much tossing about. Towards evening, cold sweating all over and became chilled; the extremities were cold and did not get warm again. Further rigors during the night; extremities still would not get warm; no sleep and some delirium which quickly passed off.

Eighth day: about noon, she became warm, thirsty and comatose; nausea, vomited a small quantity of yellowish bile-stained material. An uneasy night without sleep. Frequently unconsciously incontinent of large quantities of urine.

Ninth day: all symptoms abated; comatose. In the afternoon had rigors, vomited a small amount of bilious material.

Tenth day: rigor, fever increased in a paroxysm; had no sleep at all. Early in the morning she passed a large quantity of urine which did not show a sediment. Extremities became warm.

Eleventh day: vomited bilious rust-coloured material. Shortly afterwards, she had rigors and the extremities became cold again. Towards evening, sweating, rigors and much vomiting; a distressed night.

Twelfth day: vomited much dark, foul-smelling matter; much sobbing, a distressing thirst.

Thirteenth day: vomited much dark foul-smelling matter; rigor. About midday she lost her voice.

Fourteenth day: epistaxis. Death.

This patient throughout had relaxed bowels and shivery attacks; age about seventeen.

2. The year was rainy and southerly; throughout there was no wind. Droughts having occurred immediately before, about the rising of Arcturus, there was much rain accompanied by southerly winds. The autumn was overcast and cloudy with a very heavy rainfall. The winter was southerly and wet; mild after the solstice. Much later, near the equinox, belated storms occurred and, right at the equinox, a spell of northerly winds bringing snow but this did not last long. The spring again was southerly and calm; rainfall continued to be heavy until the rising of the Dog Star. The summer was fine and warm, and there were periods of stifling heat. The etesian winds blew feebly and at scattered intervals. Again about the rising of Arcturus, northerly winds brought much rain.

The whole year then being southerly, wet and mild, health was good during the winter, except in the case of the consumptive, about whom I shall write.

3. Early in the spring, just at the time the cold snaps occurred, there was a lot of severe erysipelas; in some cases from some obvious cause, but in others from none. Many cases proved fatal and many had a painful throat. The symptoms were a weakened voice, *causus* accompanied by brain fever, aphthae in the mouth, tumours in the pudendal region, ophthalmia, carbuncles, disorders of the bowels, loss of appetite, sometimes thirst, abnormalities of the urine which was abundant and bad. The patients were mainly comatose, but again there were periods of wakefulness. Very often there was

no crisis or it was attained with difficulty. There was also dropsy and much consumption. Such were the epidemic diseases; the sick fell into the classes given above, and many of them died. The course of the various diseases was as follows.

4. In many cases erysipelas occurred which spread all over the body on any chance happening and especially following a slight wound; those about sixty years of age were particularly liable to it in the head if any wound there were slightly neglected. Many cases, too, under treatment suffered from extensive inflammation, the erysipelas spreading rapidly in all directions. In the majority of cases abscessions turned to collections of pus. There was much destruction of flesh, sinews and bones. The fluid which formed in the abscess was not like [ordinary] pus but a different sort of morbid fluid, being both copious and varied. In those cases where something of this sort affected the head, the whole head would become bald including the beard; the bones became thin and portions became detached, while there was a discharge at many points.

These symptoms occurred both with and without fever. They were more frightening than serious, for when the disease resulted in the formation of a localized collection of pus, or some similar ripe condition, the majority recovered. On the other hand, when the inflammation and the erysipelas departed without causing such abscess formation, many sufferers died. Much the same happened if it wandered off to any other part of the body, for many had the whole arm or forearm waste away. Those whose sides were attacked, suffered harm in some part either in front or at the back of the body. In some cases, the whole thigh or the calf became thin and the whole foot too. The worst of all was if the disease attacked the pubes and private parts.

All these things happened as the result of a wound or of some obvious cause. But in many other cases it accompanied, preceded or followed fevers. In these cases, whenever localization took place with the formation of a collection of pus, or an opportune disturbance of the bowels, or favourable urine was passed, this resolved the disease. But when none of these symptoms occurred and the disease departed without giving a sign, it was fatal. By far the largest number of cases of erysipelas occurred in the spring; but it continued throughout the summer and on into the autumn.

5. Some people were very ill with swellings in the throat, inflammation of the tongue and abscesses in connection with the teeth.

A common sign at the onset, not only of consumption, but also in cases of *causus* and brain-fever, was a weakening and choking of the voice.

6. Cases of *causus* and brain-fever began early in the spring after the cold spells had passed, and many people were taken ill at that time. In these cases, the disease was acute and liable to prove fatal. The symptoms found in *causus* were as follows. To start with, the patients were comatose and nauseated, shivering and with high fever, but they were neither excessively thirsty nor delirious. Slight epistaxis occurred. In the majority of cases, the paroxysms took place on even days, and about the time of these paroxysms, there was loss of memory, exhaustion and loss of voice. The feet and hands were rather cold all the time but especially so at the paroxysms. Subsequently, the patients would get warm again slowly, but not thoroughly; they also became lucid and talked. They were also afflicted with a continuous coma but something that differed from sleep, or with a painful insomnia.

In most of these cases, the bowels were disordered, the stools being thin, raw and copious. The urine was copious too and thin, but giving none of the signs of crisis, nor any other helpful sign. In fact, those who were then attacked showed no crisis at all; there was no beneficent haemorrhage nor did critical abscess-formation of the usual sort occur. Many died after no fixed interval, but just as matters chanced; some at the crisis, some after a long period of loss of voice, some in bouts of sweating. These were the symptoms in the fatal cases, but they were much the same in brain-fever. All these cases showed a complete absence of the symptoms of thirst; nor did any of those with brain-fever go mad as in other cases, but they perished with their heads weighed down by a growing stupor.

7. There were also other fevers which I shall describe. Many had aphthae and ulcers in the mouth; many had discharges around the pudendal area; while sores and tumours both external and internal occurred, some about the groin. Moist ophthalmia occurred which was both chronic and painful. Excrescence on the eyelids, both internal and external, occurred and, in many cases, impaired the vision: the name 'figs' is given to these. There were many cases of growth on other ulcers and on the pudenda. Carbuncles were common during the summer and other septic lesions and large pustules. Many suffered from extensive herpetic lesions.

8. Frequent and dangerous disorders affecting the belly were common. First, many had distressing tenesmus; most of these were children, including all below the age of puberty, and most of these died. Many had enteritis or dysentery, but in these cases without overmuch distress. In some cases the stools were bilious, fatty, thin and watery. In many cases this was so at the inception of the disease, both with and without fever. Painful colic and malignant flatulent colic also occurred; in these going to stool did not relieve the pains, the stools being such that much remained within the bowel after attempted evacuation. This condition responded only with difficulty to medicines, and in most cases purgatives did additional harm. Many of those with this complaint perished soon; others lasted rather longer.

To sum up, whether their illnesses were long or short, all who suffered from disease of the belly were specially likely to die, for disease of the belly was a contributory factor in all the fatal cases.

9. In addition to all the previously mentioned symptoms, all suffered from loss of appetite, and that to an extent which I have never previously encountered. Those just described were especially affected, and particularly the hopeless cases both in this group and in the others mentioned. Some had a thirst, but not all. Those who had fevers or one of the other diseases

had an intemperate thirst in no case, but they would take as much or as little to drink as you wished.

10. The amount of urine passed was great; it was not proportional to the amount drunk but considerably in excess. The urine which was passed was also markedly bad, for it possessed neither thickness nor ripeness. In most cases these signs signified some wasting and disorder of the bowels with pain and no crisis.

11. Those who suffered from brain-fever and *causus* were particularly liable to become comatose, but this also occurred as an additional symptom in the other diseases in all the most serious cases, provided they were accompanied by fever. Throughout, most patients suffered either from deep coma or had only short periods of light sleep.

12. Many other types of fever were epidemic: tertians, quartans, nocturnal fevers, continued fevers, long fevers, irregular fevers, fever accompanied by nausea, and unstable fevers. All these were accompanied by much disturbance; the bowels were disordered and the patients were liable to shivering attacks. Sweating took place but did not mark the crisis; the condition of the urine has been described. In most of these cases, the illness was prolonged, for even when abscess formation did occur, it did not bring about a crisis in the way usual with other cases. In general, the diseases reached a crisis with difficulty, or there was no crisis and the illness remained chronic; this was specially the case with these people. A few of them had a crisis about the eightieth day, but in most the disease departed at no fixed time. A few died of dropsy without having been confined to their beds. Many were troubled with swellings in addition to their other diseases; especially the consumptive.

13. But it was consumption which proved the most widespread and the most serious complaint and this was responsible for most of the deaths. In many cases it began during the winter and, though many took to their beds, some of those who were ill did not do so. By early spring, most of those who had taken to their beds had died. In other cases, the cough, although it did not go away altogether, was less

troublesome during the summer. Towards autumn, all took to their beds and many died. Of these, the majority had had a long illness.

In most cases, the illness started with sudden deterioration. The symptoms were: frequent shivering attacks, often high continued fever, much untimely sweating although the patients remained cold throughout, and much chilling so that it was difficult to get them warm again. Their bowels were inconstant, constipation rapidly giving way to diarrhoea, while near the end, diarrhoea was violent in all cases. The lungs were evacuated downwards; although the urine was large in quantity it was unfavourable. Wasting was pernicious. Coughing continued throughout the illness, and it was common for patients to bring up large amounts of ripe moist sputum, without excessive pain. But even in the cases where there was pain, the process of ridding the lungs of matter took place quite mildly. The pharynx was not painful, nor did salty humours cause any trouble. There were however copious discharges from the head of sticky, white, moist and frothy material. These patients, like those already described above, suffered by far the greatest harm from their loss of appetite. They would not even take fluid nourishment, but remained without thirst. As death approached, they showed heaviness of the body, coma, swelling becoming dropsical, shivering and delirium.

14. The appearance which characterizes consumptives is a smoothness of the skin, slight pallor, freckles, a slight flush, sparkling eyes, white phlegm and winging of the shoulder-blades. The signs were the same in women too. They also show melancholy and suffused cheeks.

Causus, brain-fever and dysentery might follow upon these symptoms. The young, who were liable to phlegm, suffered from tenesmus. Those subject to bitter bile had long-lasting diarrhoea and acrid, greasy stools.

15. In all the cases so far described, the spring was the worst time and most of the deaths occurred then; the summer was the easiest time and few died then. Deaths occurred again during the autumn and under the Pleiads, in most cases on the

fourth day. It seems to me that a normal summer is beneficial. For the coming of winter terminates summer diseases, and the coming of summer shifts winter diseases. All the same, considered by itself, the summer in question was not a settled one; for it suddenly turned hot, southerly and windless, but this was beneficial by being such a change from the previous weather.

SIXTEEN CASES

17* (i)

At Thasos, the man from Paros who lay beyond the temple of Artemis took a high fever, at first of the continued type like that of *causus*, with thirst. At first he was comatose, then wakeful again; the bowels were disordered at first and the urine thin.

Sixth day: passed oily urine; delirious.

Seventh day: all symptoms more pronounced; did not sleep at all, urine unchanged, mind disordered. The stools were bilious and greasy.

Eighth day: slight epistaxis; vomited a small quantity of rust-coloured matter. Small amount of sleep.

Ninth day: no change.

Tenth day: all symptoms showed a decrease in severity.

Eleventh day: sweated all over and became chilled, but quickly got warm again.

Fourteenth day: high fever, stools bilious, thin and copious; urine contained suspended matter. Delirium.

Seventeenth day: distressed, for the patient was sleepless and the fever increased.

Twentieth day: sweating all over, no fever, stools bilious, no appetite, comatose.

Twenty-fourth day: a relapse.

Thirty-fourth day: no fever; bowels not constipated. Temperature rose again.

* Section 16 appears to be an interpolation and is omitted.

Fortieth day: no fever, bowels constipated for a short while, no appetite, slight fever returned and was, throughout, irregular; at times he was without fever at others not. Any remission and improvement was followed quickly by a relapse. He took but little food and that poor stuff. He slept badly and showed delirium about the time of the relapses. At these times he passed thicker urine, but it was disturbed and bad. The bowels were sometimes constipated, sometimes relaxed. Slight fever continued throughout, and the stools were thin and copious.

Died in 120 days.

In this case, the bowels from the first day were either loose with copious bilious stools, or he passed frothy undigested constipated stools. The urine was bad throughout. He was comatose for the most part, but he could not sleep when in pain. At no time did he have any appetite.

(ii)

At Thasos, a woman who lay near the cold spring gave birth to a daughter. The lochia were withheld and on the third day she had a high fever with shivering. For a long while before delivery she had been febrile, had kept to her bed and suffered from loss of appetite. After the preliminary rigor, fever was high and continued and accompanied by shivering. On the eighth day and subsequent days there was much delirium, but she quickly became lucid again. The bowels were disordered and she passed thin copious stools like watery bile. No thirst.

Eleventh day: was both lucid and comatose. Passed much thin dark urine. Sleepless.

Twentieth day: slight chilling but quickly became warm again; slight delirious talking, insomnia. No change in the condition of the bowels; much watery urine.

Twenty-seventh day: no fever, bowels constipated. Not long afterwards she had a violent pain in the right hip which lasted a long time. Fevers followed again; urine watery.

Fortieth day: pain in the hip lessened, but she had continued moist cough. Bowels constipated, no appetite, no change in

the urine. The fever showed no remission generally, but exacerbation occurred in no regular pattern.

Sixtieth day: the cough ceased without any signs, for there was no ripening of the sputum nor any of the other signs of localization. The mandible was in spasm and protruded to the right and the patient comatose; then followed delirious talking quickly giving way to lucidity. She was obstinately averse to food. The jaw became normal again, but she continued to pass small quantities of bilious matter in the stools. The fever was higher and was accompanied by shivering. The voice was lost in subsequent days but was regained.

Died on the eighteenth day.

In this case the urine was dark, thin and watery throughout. Coma set in and there was loss of appetite, despondency, insomnia, fits of anger and agitation associated with a melancholy disposition.

(iii)

At Thasos, Pythion, who lay beyond the temple of Heracles, had a violent rigor and high fever as the result of strain, exhaustion and insufficient attention to his diet. Tongue parched, he was thirsty and bilious and did not sleep. Urine rather dark containing suspended matter which did not settle.

Second day: about midday, chilling of the extremities, particularly about the hands and head, showed loss of both speech and voice, and he was also short of breath for a long time. Then he became warm again and thirsty. A quiet night; slight sweating about the head.

Third day: quiet. Late in the day, about sunset, slight chilling, nausea, disturbed bowels followed by an uneasy sleepless night. Passed a small constipated stool.

Fourth day: morning quiet. About midday, all symptoms more pronounced; chilled, aphasia, aphonia became worse. After a while he became warm again and passed dark urine containing suspended matter. A quiet night; slept.

Fifth day: seemed to improve, but there was a painful heaviness in the belly. Thirsty. An uneasy night.

Sixth day: morning quiet. In the evening the pains were more severe and paroxysmal. The bowels were well opened late at night after an enema. Slept during the night.

Seventh day: nausea, somewhat distressed. Passed oily urine. Much disturbed during the night, random talking, no sleep.

Eighth day: slept a little in the morning, but soon became chilly with loss of voice. Respiration was superficial and shallow, late at night he became warm again, but delirious. A slight improvement took place towards daybreak. Stools unmixed, small and bilious.

Ninth day: comatose; nauseated whenever he woke. No excessive thirst. About sunset he became distressed, talked at random and this was followed by a bad night.

Tenth day: in the morning, he lost his voice, became very chilled, had high fever with much sweating and died.

In this case the distress was marked on the even days.

(iv)

A patient with brain-fever took to his bed on the first day of the illness and vomited much rust-coloured thin matter. He had severe fever accompanied by shivering and continuous sweating of the whole body. There was a painful heaviness of the head and neck. The urine was thin and contained a small amount of scattered particles suspended in it, but did not sediment. He passed a large single stool, became delirious and did not sleep.

Second day: he was voiceless in the morning; fever high, sweating without remission. The whole body throbbed and convulsions occurred during the night.

Third day: all symptoms more pronounced.

Fourth day: died.

(v)

At Larisa, a bald man suddenly had a pain in the right thigh. No treatment which he received did him any good.

First day: high fever of *causus* type, did not tremble, but the pain continued.

Second day: pains in the thigh were relieved, but the fever increased. The patient became somewhat distressed and did not sleep; the extremities were cold. He passed a lot of urine but this was not of a favourable kind.

Third day: the pain in the thigh ceased. His mind became unhinged and there was much disturbance and tossing about.

Fourth day: died about noon.

(vi)

At Abdera, Pericles took a high fever of continued type, accompanied by distress. He had much thirst, was nauseated and could not keep liquids down. The spleen was enlarged and he had headache.

First day: epistaxis from the left nostril; the fever however increased considerably. He passed much cloudy white urine which did not sediment on standing.

Second day: all symptoms more pronounced. The urine however was thick and settled more. The nausea was less severe and the patient slept.

Third day: the fever became less and he passed a large quantity of ripe urine with a lot of sediment. A quiet night.

Fourth day: about noon he had a warm sweat involving the whole body, the fever left him and he reached the crisis. There was no relapse.

(vii)

A girl who lay at a house on the Sacred Way at Abdera took a fever of the *causus* type. She complained of thirst and was wakeful. Menstruation took place for the first time.

Sixth day: much nausea, redness and shivering; she was distraught.

Seventh day: no change. The urine, though thin, was of good colour and there was no trouble with the bowels.

Eighth day: deafness supervened, with high fever, insomnia, nausea and shivering. She became lucid. Urine the same.

Ninth day: no change, nor on the following days. The deafness persisted.

Fourteenth day: mind disordered; the fever became less.

Seventeenth day: a large epistaxis; the deafness became slightly less. Nausea and deafness were present on the following days as well as some delirium.

Twentieth day: pain in the feet and deafness. The delirium ceased and there was a slight nose-bleed and sweating; no fever.

Twenty-fourth day: the fever returned and she was deaf again. The pain in the feet continued and her mind was wandering.

Twenty-seventh day: severe sweating and lost her fever. The deafness cleared up and, although pain in the feet remained, in other respects the final crisis was reached.

(viii)

Anaxion, who lay near the Thracian Gates at Abdera, took a high fever. There was continuous aching in the right side and a dry cough, but no spitting in the first few days. He suffered from thirst and insomnia, but the urine was copious, thin and of good colour.

Sixth day: delirium. No improvement as the result of warm fomentations.

Seventh day: distressed as the fever increased. The pain did not decrease, cough was troublesome and breathing difficult.

Eighth day: I bled him at the elbow; there was a large flow of blood as there should be. The pain decreased but the dry cough continued.

Eleventh day: the fever decreased. There was slight sweating about the head, while the cough and the sputum from the lungs were moister.

Seventeenth day: began to expectorate a small quantity of ripe sputum and his condition improved.

Twentieth day: sweated and lost his fever. After the crisis he was thirsty and the matter evacuated from the lungs was not good.

Twenty-seventh day: the fever returned and, with coughing, he brought up much ripe matter. A large white sediment in the urine. His thirst was lost and his respiration became normal.

Thirty-fourth day: sweated all over; no fever. A complete crisis.

(ix)

Heropythus at Abdera had a headache; he remained up for a while but eventually went to bed with it. He lived near the upper highroad. He showed a fever of the *causus* type. At first he vomited much bilious matter and suffered from thirst and much distress. His urine was thin and dark; sometimes, but not always, it contained suspended matter. An uneasy night. The fever showed paroxysms at varying intervals, for the most part quite irregular.

About the fourteenth day, he complained of deafness and the fever increased; the urine remained as before.

Twentieth and following days: much delirium.

Fortieth day: a large epistaxis and became more lucid. The deafness was still present but was less severe. The fever abated. Small epistaxes occurred frequently on the following days.

About the sixtieth day the haemorrhages stopped but there was a violent ache in the right hip and the fever increased again. A little later, there was pain involving all the lower part of the body. It so happened that either the temperature was up and the deafness worse, or these two symptoms abated while the pain in the lower part of the body and about the hips became worse.

From the eightieth day onwards all the symptoms decreased, although none entirely disappeared. He passed urine of good colour with more sediment.

About the hundredth day, there was disorder of the bowels with the passage of copious bilious stools. This went on to a considerable extent for not a little while. The signs of dysentery, accompanied by pain, were associated with an easing off of the other symptoms. Generally speaking, the fever departed and the deafness ceased.

A final crisis took place on the 120th day.

(x)

Nicodemus took a fever at Abdera as the result of sexual indulgence and drinking. To start with he suffered from nausea and pain in the heart, thirst and a parched tongue. His urine was thin and dark.

Second day: paroxysms of fever, shivering, nausea; no sleep. Vomited yellow bilious matter. Urine as before. A quiet night and he slept.

Third day: a general remission and improvement. About sunset he became somewhat distressed again and passed an uneasy night.

Fourth day: a rigor, much fever and pains all over. Urine thin containing suspended matter. Night again quiet.

Fifth day: all symptoms continued but were less pronounced.

Sixth day: pain all over as before; the urine contained suspended matter. Delirium.

Seventh day: improved.

Eighth day: all other symptoms abated.

Tenth and following days: the pains continued but were all less acute. The paroxysms and pain in this case were throughout more pronounced on the even days of the illness.

Twentieth day: passed white urine which, although thick, did not form a sediment on standing. Much sweating; appeared to lose his fever but again became warm in the evening with pains as before, shivering, thirst and slight delirium.

Twenty-fourth day: passed much white urine which contained a large quantity of sediment. Sweated all over profusely, the sweat being warm. Lost his fever as the crisis was passed.

(xi)

A woman at Thasos became morose as the result of a justifiable grief, and although she did not take to her bed, she suffered from insomnia, lose of appetite, thirst and nausea. She lived on the level ground near Pylades' place.

Early on the night of the first day, she complained of fears and talked much; she showed despondency and a very slight fever. In the morning she had many convulsions; when the convulsions had for the most part ceased, she talked at random and used foul language. Many intense and continuous pains.

Second day: condition unchanged; no sleep and the fever higher.

Third day: the convulsions ceased but lethargy and coma supervened followed by a return to consciousness, when she leapt up and could not be restrained. There was much random talking and high fever. That night she sweated profusely all over with warm sweat. She lost her fever and slept, becoming quite lucid and reaching the crisis.

About the third day, the urine was dark and thin, and contained suspended matter, for the most part round particles, which did not sediment. About the time of the crisis, a copious menstrual discharge took place.

(xii)

A girl at Larisa took a high fever of the *causus* type. She had insomnia and thirst while her tongue was dry and smoke-coloured. The urine was of good colour but thin.

Second day: distressed; did not sleep.

Third day: the stools were bulky, watery and greenish. On the following days stools of similar character were passed without distress.

Fourth day: passed a small quantity of thin urine which contained suspended matter which did not settle. Delirium during the night.

Sixth day: a violent and copious epistaxis. Shivering was followed by profuse hot sweating all over; she lost her fever and reached the crisis.

She menstruated for the first time during this illness, while the fever was still present, but after the crisis. She was only a girl. Throughout, she suffered from nausea, shivering, a flushed face, aching eyes and heaviness of the head. In this case there was no relapse but a single crisis. The distress was experienced on the even days.

(xiii)

Apollonius at Abdera suffered for a long time without taking to his bed. He had an enlarged abdomen and a pain in the region of the liver to which he had become accustomed, for he became jaundiced, flatulent and of pallid complexion.

As a result of eating beef and drinking cows' milk intemperately, he developed what was a slight fever at first and went to bed. He got much worse through taking a large amount of milk, both boiled and cold, both goats' and sheep's, and by taking a generally bad diet. For the fever increased and he passed nothing worth mentioning in the stools of the food he took. He passed little urine and that was thin. He was unable to sleep.

He then became badly distended, suffered from thirst and became comatose. There was swelling, accompanied by an aching pain in the right hypochondrium. All the extremities were somewhat cold. He began talking at random, showed loss of memory in anything he said, and became disorientated.

About the fourteenth day from the time he took to his bed he had rigors, his temperature rose and he went out of his mind; there was shouting, disturbance and much talking, then he settled down again and relapsed into coma. Subsequently his bowels were upset, the stools being copious, bilious, raw and unmixed. The urine was dark, small in quantity and thin. There was much distress. The excreta were not always the same; sometimes they were small in quantity and dark and rust-coloured, or they were greasy, raw and pungent. At times too he seemed to pass milky substances.

About the twenty-fourth day he was more comfortable; in other respects the symptoms were unchanged, but he became slightly lucid. He could remember nothing from the time he took to his bed. Shortly afterwards, his mind was again disordered and there was a general tendency to deteriorate.

About the thirtieth day he had high fever, copious thin stools and delirium. Extremities cold; loss of voice.

Thirty-fourth day: died.

Throughout this case, from the time I knew of it, the bowels were disordered and the urine was thin and dark; the patient also suffered from coma, insomnia and cold extremities, and he was delirious throughout.

(xiv)

At Cyzicus, a woman gave birth to twin girls; the labour was difficult and the lochia abnormal.

On the first day, there was high fever with shivering, and heaviness and aching of the head and neck. She was sleepless from the start and she was silent, scowling and disobedient. Urine thin and of bad colour, thirst, nausea for the most part, diarrhoea and constipation succeeding each other at no fixed intervals.

Sixth day: much random talking during the night; no sleep.

About the eleventh day went mad and then became lucid again; urine dark, thin and then, after an interval, oily. The bowels were disturbed, the stools being large in quantity and thin in consistency.

Fourteenth day: many convulsions, extremities cold, still no trace of lucidity, suppression of urine.

Sixteenth day: loss of voice.

Seventeenth day: died.

(xv)

At Thasos the wife of Delearces, who lay on the level ground, took a high fever with shivering as the result of grief. From the start she used to wrap herself up, always remaining silent while she groped about, scratching and plucking out hair, and alternately wept and laughed. She did not sleep. She remained constipated even when the bowels were stimulated. She drank a little when reminded to do so; the urine was thin and small in quantity. Fever was slight to the touch; the extremities were chilly.

Ninth day: much random talking, but subsequently she quietened down and fell silent.

Fourteenth day: respiration infrequent; deep for a while and then the breaths would be short.

Seventeenth day: the bowels were stimulated and disordered stools were passed giving way to the actual liquid drunk, nothing being retained. She was insensible to everything. Skin taut and dry.

Twentieth day: much talking and then quietened down again; loss of voice, respiration in short breaths.

Twenty-first day: died.

Throughout this case respiration was intermittent, and deep. She was insensible to everything, always kept herself wrapped up and either talked at random or kept silence.

(xvi)

At Meliboea, a young man who had been running a temperature for a long time as the result of drinking and much sexual indulgence, took to his bed. His symptoms were shivering, nausea, insomnia and lack of thirst.

On the first day, his bowels passed a large quantity of solid faeces accompanied by much fluid. On the following days he passed a large quantity of watery, greenish stools. His urine was thin and small in quantity, and of bad colour. Respiration at long intervals and deep after a while. There was a somewhat flabby distension of the upper part of the abdomen extending laterally towards the flanks. Palpitation of the heart was continuous throughout. He passed oily urine.

Tenth day: was delirious without excitement, being well-behaved and silent. Skin dry and taut. Stools either copious and thin, or bilious and greasy.

Fourteenth day: all symptoms more pronounced. Delirium with much talking at random.

Twentieth day: went mad, much tossing about, passed no urine, kept down a small amount of fluid.

Twenty-fourth day: died.

THE SCIENCE OF MEDICINE

This spirited defence of Medicine is a remarkable document of an age when there were no precautions against unqualified practitioners, and all physicians were exposed to charges of charlatanry. The title is often rendered The Art, *but this gives the wrong impression, for it is the writer's main contention that Medicine is an exact science, not an undefinable art.*

1. There are men who have turned the abuse of the arts and sciences into an art in itself and, although they would not confess it themselves, their aim nevertheless is simply to display their own knowledge. But it seems to me that it is the aim and function of an intelligent mind to make new discoveries in whatever field such investigations may be useful, and also to bring to completion tasks that are but half-finished. On the other hand, a desire to use the art of abuse to belittle the scientific discoveries of others and to slander the discoveries of the learned to the illiterate, rather than to offer constructive criticism, is not so much the aim and function of an intelligent mind, as a proof of warped character and want of skill. Those who have the ambition to be scientists but not the necessary ability are equipped for the malicious habit of slandering their neighbours' work if it be right, or of censuring it if it be wrong. In other sciences, let those who can stop their enemies, each in his own subject. This thesis aims at answering the opponents of the science of medicine, deriving boldness from the character of those it censures, facility from the subject it defends and strength from its trained judgement.

2. It appears to me that there is no science which has no basis in fact. It would be absurd to suppose something that exists non-existent. For what being could anyone ascribe to a non-existent thing as a proof of its existence? If it were possible to see what has no substance, just as we see what does exist, then one could no longer call such a thing non-existent because it would then appear alike to the eye and the mind existent. But may not the truth be something like this: what

exists is always visible and recognizable, and what does not exist is neither visible nor recognizable? The activities of the sciences that are taught are things that can be seen and there is none that is not visible in one form or another. I at least am of the opinion that it is from the visible forms of things that they take their names. It is absurd to suppose that forms spring from names; that were impossible since names are adopted by convention, whereas forms are not invented but are characteristic of those things from which they spring.

3. If some of my readers have not sufficiently grasped the argument, it may be explained more clearly in other words. Let us consider the science of medicine, since that is my own subject, by way of illustration. First of all I would define medicine as the complete removal of the distress of the sick, the alleviation of the more violent diseases and the refusal to undertake to cure cases in which the disease has already won the mastery, knowing that everything is not possible to medicine. It is my intention to prove that medicine does accomplish these things and is ever capable of doing them. And as I describe the science I shall at the same time disprove the arguments of her traducers, whatever way each prides himself on his attack.

4. My first premise is one that everyone accepts; for it is admitted that some who have received medical attention have been restored to health. But the fact that everyone is not cured is reckoned an argument against the science, while those who recover from their diseases, so the traducers of the science assert, owe their cure to good fortune rather than to medical skill. Even I do not exclude the operations of fortune, but I think that those who receive bad attention usually have bad luck, and those who have good attention good luck. Secondly, what else but medical skill can be responsible for the cures of patients when they have received medical attention? Such, not content to wait on the shadowy form of Fortune, entrusted themselves to the science of medicine. While the share of chance is excluded, that of science is not. They submitted themselves to its ordinances and they had faith in it; they

considered its apparent nature and the result proved to them its effectiveness.

5. My opponents will say that many sick men have never seen a doctor and yet have recovered from their illnesses. I do not doubt it. But it seems to me that even those who do not employ a doctor may chance upon some remedy without knowing the right and wrong of it. Should they be successful, it is because they have employed the same remedy as a doctor would use. And this is a considerable demonstration of the reality and the greatness of the science, when it be realized that even those who do not believe in it are nevertheless saved by it. For when those who employ no doctors fall sick and then recover, they must know that their cure is due either to doing something or to not doing it. It may be fasting or eating a great deal, drinking largely or taking little fluid, bathing or not bathing, exercise or rest, sleep or wakefulness, or perhaps it is a mixture of several of these that is responsible for their cure. If they benefit, they cannot help but know what benefited them; if they are harmed, what harmed them; but everyone cannot tell what is going to bring benefit or harm beforehand. If a sick man comes to praise or to blame the remedies by which he is cured, he is employing the science of medicine. The failure of remedies too is no less a proof of the reality of the science. Remedies are beneficial only through correct applications, but they are harmful when applied wrongly. Where there are procedures which can be right or wrong, a consideration of these must constitute a science. I assert that there is no science where there is neither a right way nor a wrong way, but science consists in the discrimination between different procedures.

6. If the science of medicine and the profession were concerned in their cures only with the administration of drugs, purges and their opposites, my argument were a weak one. But the most renowned physicians are to be seen employing as therapeutic measures, diets and other ordinances which not even an untaught layman, much less a doctor, could deny were part of their science. There is nothing done which is useless by good doctors, nor is there anything useless in the science of

medicine. The majority of plants and preparations contain substances of a remedial or pharmaceutical nature and no one who is cured without the services of a doctor can ascribe his cure to chance. Indeed, upon examination, the reality of chance disappears. Every phenomenon will be found to have some cause, and if it has a cause, chance can be no more than an empty name. The science of medicine is seen to be real both in the causes of the various phenomena which occur and in the provisions which it takes to meet them, nor will it ever cease to be so.

7. This will suffice as a reply to those who, to the disparagement of medicine, attribute their health to luck. But those who use the example of patients who die from their illnesses as an argument against the efficacy of medicine make me wonder what trustworthy reason leads them to absolve a patient's weakness of character, and impute instead a lack of intelligence on the part of his physician. As if doctors can prescribe the wrong remedies but patients can never disobey their orders! It is far more likely that the sick are unable to carry out the instructions than that the doctors prescribe the wrong remedies. Physicians come to a case in full health of body and mind. They compare the present symptoms of the patient with similar cases they have seen in the past, so that they can say how cures were effected then. But consider the view of the patients. They do not know what they are suffering from, nor why they are suffering from it, nor what will succeed their present symptoms. Nor have they experience of the course of similar cases. Their present pains are increased by fears for the future. They are full of disease and starved of nourishment; they prefer an immediate alleviation of pain to a remedy that will return them to health. Although they have no wish to die, they have not the courage to be patient. Such is their condition when they receive the physician's orders. Which then is more likely? That they will carry out the doctor's orders or do something else? Is it not more likely that they will disobey their doctors rather than that the doctors, whose attitude I have outlined above, will prescribe the wrong remedies? There can be no doubt that the patients are likely to

be unable to obey and, by their disobedience, bring about their own deaths. So they are wrong who attribute the blame to the innocent and exculpate the guilty.

8. There are some too who condemn the science of medicine because doctors are unwilling to tackle incurable cases. They allege that such diseases as the physicians do attempt to treat would get better of themselves in any case, while those that need medical attention are neglected. If medicine were really a science, they say, all should be cured alike. In truth, this accusation would be a better one if they blamed the doctors for neglecting to treat such madmen as themselves. A man who thinks that a science can perform what is outside its province, or that nature can accomplish unnatural things, is guilty of ignorance more akin to madness than to lack of learning. Our practice is limited by the instruments made available by Nature or by Art. When a man is attacked by a disease more powerful than the instruments of medicine, it must not be expected that medicine should prove victorious. For example, fire is the most powerful caustic known to medicine, although there are many other caustics employed which are less powerful. Now it is not reasonable to call a disease incurable because it does not yield to the weaker remedies, but if it does not respond to the most powerful measures, it is clearly incurable. When fire fails to produce some particular effect, is it not plain that such an effect can only be accomplished by some science other than that whose instrument fire is? The same argument holds good for the failure of other methods employed in medical practice. For these reasons, then, I assert that when the physicians fail, it is the power of the disease which is responsible and not deficiencies in the science of medicine. These critics, then, would have us spend as much time on incurable patients as on those we can do something for. Thus do they impress doctors who are physicians in name alone while they are a laughing-stock to the genuine practitioners of the science. Experts in their professions require neither praise nor blame from such fools as they. Rather do they want criticism from men who have considered what are the full services medicine can

render, what deficiencies remain and, if there are deficiencies, how much must be attributed to the failure of the physicians and how much to the patient.

9. The defence of the other sciences must await another time and another book. Here I am concerned only with the science of medicine, to discuss its nature and how its practice should be judged. I have already demonstrated this in part; I now proceed to the rest. Those who are reasonably proficient in the science of medicine can distinguish two classes of disease. There is a small group in which the signs are readily seen by the eye, those in which the flesh is changed in appearance or in which swellings are demonstrable. Then there is a large group not so easily diagnosed. In the former group, signs can be elicited by sight and touch, for instance, whether the skin be firm or clammy, hot or cold, and each of such signs is of significance. In this group of diseases cure should be complete, not because such diseases are necessarily more amenable to treatment, but simply because the cure has been discovered. And the discoveries were made by no chance comers, but by experts who were in a position to make them. However, anyone can be trained to be an expert who is lacking neither in education nor intelligence.

10. There should be no difficulty, then, in the management of these obvious or external diseases, but the less obvious or internal diseases should not be wholly beyond the power of the science. In this latter group I include disease of the bones and of the cavities of the body. The body contains many hollow organs; there are two which receive and pass on the food and many others which those who have studied them will know. Every part of the body which is covered with flesh or muscle contains a cavity. Every separate organ, whether covered by skin or muscle, is hollow, and in health is filled with life-giving spirit; in sickness it is pervaded by unhealthy humours. The arms, for example, possess such a cavity, as also do the thighs and legs. Even those parts which are relatively poorly covered with flesh contain such cavities. Thus the trunk is hollow and contains the liver, the skull contains the brain and the thorax the lungs. Thus the divisions

of the body may be likened to a series of vessels, each containing within it various organs, some of which are harmful and some beneficial to their possessor. There are in addition many blood-vessels and nerves which do not lie loose among the muscles but are attached to the bones and ligaments which form the joints. The joints themselves, in which the ends of the bones turn, are enclosed by capsules which contain a frothy fluid. Should the joint be opened, large quantities of fluid escape and much damage is done.

11. Since these diseases cannot be diagnosed by sight, I call them 'internal diseases' and such is the term employed by the profession. These internal diseases have not been mastered, but they have been mastered as far as possible for the present. The future depends on how far the intelligence of the patients permits the drawing of conclusions and how far the abilities of future investigators are fitted for the task. If the nature of a disease cannot be perceived by the eye, its diagnosis will involve more trouble and certainly more time than if it can. What escapes our vision we must grasp by mental sight, and the physician, being unable to see the nature of the disease nor to be told of it, must have recourse to reasoning from the symptoms with which he is presented. Then when sick men suffer from delay in diagnosis, it is due rather to the nature of the disease and of the patient than to the failure of the physician. It is made more difficult by the fact that the symptoms which patients with internal diseases describe to their physicians are based on guesses about a possible cause rather than knowledge about it. If they knew what caused their sickness they would know how to prevent it. To know the cause of a disease and to understand the use of the various methods by which disease may be prevented amounts to the same thing in effect as being able to cure the malady. When the physician cannot make an exact diagnosis from the patient's description of his symptoms, the doctor must employ other methods for his guidance and any delay in diagnosis is due to the nature of the human frame rather than to a failure of the science of medicine. Medicine aims to cure that which is perceived, treatment being based on judgement rather than on ill-

considered opinion, on energy rather than indifference. The nature of the body is such that a sickness which is clearly seen can be cured. However, a disease may progress rapidly while the diagnosis is slowly becoming apparent to the physician and the patient cannot be saved in time. The progress of a disease is never faster than the speed with which it may be cured and, so long as the administration of remedies begins with the onset of the malady, recovery may be expected. But when the disease has a start because it lurks unseen within the body, a sufferer seeks treatment not when first attacked but only after his malady has gained a firm hold.

12. Thus the efficacy of the science is better demonstrated when it succeeds in relieving an internal malady than if the cure of an apparently hopeless case should be attempted. Different principles guide other crafts. A trade in which the use of fire is necessary cannot be practised in the absence of this element. Further, other crafts are exercised on materials in which mistakes can easily be rectified, as is the case with those which employ wood or hides, or in the craft of engraving on bronze, iron or similar metal. A mistake made in the manufacture of articles from such materials is easily corrected, but the craft cannot be practised at all if one of the materials be missing. Again, the time factor is not of importance and careful workmanship produces better results than speed, although the latter may prove more profitable.*

13. But although neither deep abscesses nor diseases of the kidneys nor of the liver nor of other organs situated within the body are visible to the eye, which is the most satisfactory way of observation, medicine has none the less found out means by which a diagnosis may be reached. Such means consist of observations on the quality of the voice, whether it be clear or hoarse, on the respiratory rate, whether it be quickened or slowed, and on the constitution of the various fluids which flow from the orifices of the body, taking into account their smell and colour, as well as their thinness or viscosity. By weighing up the significance of these various signs it is possible to deduce of what disease they are the result, what

*It is possible that something has been lost from the text of this section.

has happened in the past and to prognosticate the future course of the malady. Even when nature herself does not produce such signs, they may be revealed by certain harmless measures known to those practised in the science. Thus the physicians may determine what remedies should be applied. For instance, a patient may be made to rid himself of phlegm by the administration of certain acid draughts and foods. Thus a visible sign is produced of some underlying disease which could not otherwise be demonstrated to the sight. If a patient be made to walk uphill or to run, abnormalities in respiration will be observed which would not be apparent at rest. By producing sweating in the manner just mentioned the signs of fever can be observed, just as the steam from hot water indicates fire. Substances can be given which, excreted in the urine, or through the skin, reveal the disease better. Then draughts and substances taken by mouth have been discovered, which, producing more heat than the cause of some fever, act on that cause and make the fever flow away, a result which would not happen but for the exhibition of such treatment. But both the methods to be employed and the signs produced differ from case to case. As a result the signs may be difficult for the physician to interpret and then cures are slow and mistrust in the power of the doctor persists.

14. That the science of medicine makes use of principles which can be of real assistance has been shown in this work. But it would not be fair to expect medicine to attempt cures that are all but impossible, nor to be unfailing in its remedies. That medicine can be of value is further demonstrated by the skill of those proficient practitioners whose actions are better proof than their words. It is not that such physicians look down on writers, but they believe that most men are more ready to believe what they see than what they hear.

AIRS, WATERS, PLACES

An essay on the influence of climate, water supply and situation on health.

1. Whoever would study medicine aright must learn of the following subjects. First he must consider the effect of each of the seasons of the year and the differences between them. Secondly he must study the warm and the cold winds, both those which are common to every country and those peculiar to a particular locality. Lastly, the effect of water on the health must not be forgotten. Just as it varies in taste and when weighed, so does its effect on the body vary as well. When, therefore, a physician comes to a district previously unknown to him, he should consider both its situation and its aspect to the winds. The effect of any town upon the health of its population varies according as it faces north or south, east or west. This is of the greatest importance. Similarly, the nature of the water supply must be considered; is it marshy and soft, hard as it is when it flows from high and rocky ground, or salty with a hardness which is permanent? Then think of the soil, whether it be bare and waterless or thickly covered with vegetation and well-watered; whether in a hollow and stifling, or exposed and cold. Lastly consider the life of the inhabitants themselves; are they heavy drinkers and eaters and consequently unable to stand fatigue or, being fond of work and exercise, eat wisely but drink sparely?

2. Each of these subjects must be studied. A physician who understands them well, or at least as well as he can, could not fail to observe what diseases are important in a given locality as well as the nature of the inhabitants in general, when he first comes into a district which was unfamiliar to him. Thus he would not be at a loss to treat the diseases to which the inhabitants are liable, nor would he make mistakes as he would certainly do had he not thought about these things beforehand. With the passage of time and the change of the seasons, he

would know what epidemics to expect, both in the summer and in the winter, and what particular disadvantages threatened an individual who changed his mode of life. Being familiar with the progress of the seasons and the dates of rising and setting of the stars, he could foretell the progress of the year. Thus he would know what changes to expect in the weather and not only would he enjoy good health himself for the most part but he would be very successful in the practice of medicine. If it should be thought that this is more the business of the meteorologist, then learn that astronomy plays a very important part in medicine since the changes of the seasons produce changes in diseases.

3. I shall explain clearly the way in which each of these subjects should be considered. Let us suppose we are dealing with a district which is sheltered from northerly winds but exposed to the warm ones, those, that is, which blow from the quarter between south-east and south-west; and that these are the prevailing winds. Water will be plentiful but it will consist chiefly of brackish surface water, warm in the summer and cold in the winter. The inhabitants of such a place will thus have moist heads full of phlegm, and this, flowing down from the head, is likely to disturb their inner organs. Their constitution will usually be flabby and they tolerate neither food nor drink well. It is a general rule that men with weak heads are not great drinkers because they are particularly liable to hangovers.

The local diseases are these. The women are sickly and liable to vaginal discharges; many of them are sterile, not by nature, but as the result of disease. Miscarriages are common. Children are liable to convulsions and asthma which are regarded as divine visitations and the disease itself as 'sacred'. The men suffer from diarrhoea, dysentery, ague and, in the winter especially, from prolonged fevers. They are also subject to pustular diseases of the skin which are particularly painful at night and also from haemorrhoids. Pleurisy, pneumonia and other acute diseases are rare since such diseases do not flourish in a watery constitution. Moist ophthalmia is not uncommon, but it is neither serious nor of long duration unless an epidemic breaks out owing to some great change in the weather. Catarrh of the head makes those over fifty liable to hemiplegia. They suddenly become 'sunstruck' or cold. Such then are the diseases of the country, except that changes in the weather may produce epidemics in addition.

4. Let us now take the case of a district with the opposite situation, one sheltered from the south but with cold prevailing winds from the quarter between north-west and north-east. The water supply is hard and cold and usually brackish. The inhabitants will therefore be sturdy and lean, tend to constipation, their bowels being intractable, but their chests will move easily. They will be more troubled with bile than with phlegm; they will have sound and hard heads but suffer frequently from abscesses. The special diseases of the locality will be pleurisy and the acute diseases. This is always the case when bellies are hard. Because of this too, and because they are sinewy, abscesses commonly appear on the slightest pretext. This is also due to their dryness and the coldness of the water. Such men eat with good appetites but they drink little; one cannot both eat and drink a great deal at the same time. Ophthalmia occurs and is of long duration tending to become both serious and chronic, and the eyes suppurate at an early stage. Those under thirty suffer from epistaxis which is serious in summer. Cases of the 'sacred disease' are few but grave. These men live longer than those I described before. Ulcers do not suppurate nor do they spread wildly. Characters are fierce rather than tame. These then are the diseases to which the men of such a district are liable; others only if some change in the weather provokes an epidemic.

The women suffer largely from barrenness owing to the nature of the water; this is hard, permanently so, and cold. Menstruation, too, does not occur satisfactorily but the periods are small and painful. They give birth with difficulty but, nevertheless, miscarriages are rare. After parturition they are unable to feed their babies because the flow of milk is dried up by the intractable hardness of the water. As a result of difficult labour, abscesses and convulsions commonly occur and wasting disease follows. The children suffer from dropsy of the testicles while they are young, but this disappears as they grow up. Puberty is attained late in such a district.

5. So much for the influence of the warm and cold winds. Let us now consider districts which are exposed to winds from the quarter between north-east and south-east, and then those from the west. Those that face east are likely to be healthier than those facing north or south even if such places are only a furlong apart. These districts do not experience such extremes of heat and cold. The water, to the easterly side, must necessarily be clean, sweet-smelling, soft and pleasant. This is because the early morning sunshine distils dew from the morning mist. The inhabitants are generally of good and healthy complexion unless they are subject to disease. They have loud and clear voices and if, as is probable, local conditions generally are better, they are of better temperament and intelligence than those exposed to the north. The climate in such a district may be compared with the spring in that there are no extremes of heat and cold. As a consequence, diseases in such a district are few and not severe. In general, it may be said that they resemble districts of southern aspect except that the women are prolific and give birth easily.

6. Towns that face west and are thus sheltered from easterly winds while the warm winds and those from the south pass them by, must necessarily have a most unhealthy situation. First, the water is not clear. This is because the air holds the early morning mist and such air, mixing with water, takes away its sparkle, for it does not get the sun on it until late in the day. In summer damp breezes blow and cause dew to fall in the early morning, but for the rest of the day the sun, as it declines, burns up the inhabitants. This tends to make them of poor complexion and sickly and they suffer from all the diseases previously mentioned without exception. Their voices are thick and somewhat hoarse on account of the air which tends to be impure and unhealthy. Not even the northerly gales reach such districts to dispel these characteristics. All the winds that blow are from the west and therefore very wet.

The weather of such a district can be compared with the autumn when there is so great a difference between morning and evening.

7. So much then for the effects, both good and ill, of the various winds. Now I should like to explain what is the effect of different kinds of water, to indicate which are healthy and which unhealthy, and what effects, both good and bad, they may be expected to produce. Water plays a most important part in health. Stagnant water from marshes and lakes will necessarily be warm, thick and of an unpleasant smell in summer. Because such water is still and fed by rains, it is evaporated by the hot sun. Thus it is coloured, harmful and productive of biliousness. In winter it will be cold, icy and muddied by melting snow and ice. This makes it productive of phlegm and hoarseness. Those who drink it also have large and firm spleens while their bellies are hard, warm and thin. Their shoulders, the parts about the clavicles and their faces are thin too because their spleens dissolve their flesh. Such men have a great appetite for food and drink. Their viscera will be very dry and warm and thus require the stronger drugs. Their spleens remain enlarged summer and winter and, in addition, cases of dropsy are frequent and fatal to a high degree. The reason for this is the occurrence, during the summer, of much dysentery and diarrhoea together with prolonged quartan fevers. Such diseases, when they are of long standing, cause dropsy in people of this type and this proves fatal. These, then, are the summer ailments. In winter, the younger men are liable to pneumonia and to madness. The older men suffer from a fever called *causus* on account of the hardness of their bellies, the women from tumours and leucorrhoea. The latter are weak in the belly and give birth with difficulty. The foetus is large and swollen. During lactation, wasting and pains occur and menstruation does not become properly re-established. The children are specially liable to rupture and the men to varicose veins and ulcers of the legs. People of such nature cannot be long-lived and they become prematurely aged. Moreover, sometimes the women appear to have conceived but, when the time of birth approaches, the contents of

the belly disappear. This happens when the womb suffers from dropsy. Water which produces these things, I consider harmful in every respect.

We now come to the consideration of water from rock springs. It is hard; either from the soil containing hot waters, or from iron, copper, silver, gold, sulphur, alum, bitumen or nitre. All these substances are formed by the influence of heat. The water from such ground is bad since it is hard, heating in its effect, difficult to pass and causes constipation.

The best water comes from high ground and hills covered with earth. This is sweet and clean and, when taken with wine, but little wine is needed to make a palatable drink. Moreover, it is cool in summer and warm in winter because it comes from very deep springs. I particularly recommend water which flows towards the east, and even more that which flows towards the north-east, since it is very sparkling, sweet-smelling and light. Water that is salty, hard and cannot be softened, is not always good to drink. But there are some constitutions and some diseases which benefit by drinking such water and these I shall proceed to detail. The best type of this water is that which comes from springs facing the east. The second best from springs facing the quarter between north-east and north-west, especially the more easterly, and the third from springs between north-west and south-west. The worst is the southern variety, the springs facing between south-west and south-east. These water supplies are worse when the winds are southerly than when they are northerly.

Waters should be used in the following way. A man who is in good and robust health need not distinguish between them, but he may drink whatever is to hand at the moment. But if a sick man wishes to drink what is best for him, he would best regain his health by observing the following rule. If his stomach is hard and liable to become inflamed, the sweetest, lightest and most sparkling water is best for him; but if his stomach is soft, moist and full of phlegm, the hardest and saltiest are best since these will best dry it up. The water that is best for cooking and softest is likely to relax and soften the stomach. Hard water that is not softened by boiling tends to make the stomach contract and dries it up. Owing to ignorance, there is a general fallacy about brackish water. Salty water is thought to be a laxative; actually the opposite is the case and permanently hard water tends to make the bowels costive.

8. We now pass from spring water to a consideration of rain water and water from snow. Rain water is very sweet, very light and also very fine and sparkling, since the sun, drawing it up, naturally seizes upon the finest and lightest water, as is proved by the salt which is left behind. The brine is left on account of its thickness and heaviness and becomes salt, but the sun draws up the finest elements because of their lightness. It draws it up not only from ponds, but also from the sea and in fact from any source which contains moisture; and there is nothing that does not contain some. Even from human beings, it draws off the finest and lightest part of the body's humours. A very good proof of this is seen when a man goes and sits in the sun wearing a cloak. Where sunlight falls on the body, no sweat will be seen, but the part which is shaded or protected by something becomes damp with sweat. This is because the sun draws up the sweat and makes away with it; but where the body is shaded, the sweat remains because the sunlight cannot get at it. If the man goes in the shade, the whole body sweats alike because the sun is no longer on him. Rain water, being composed of a mixture of so many elements, quickly becomes rotten on standing and exhales a foul smell. But when it has been drawn up into the air, it travels round and mixes with the air; the dark and cloudy part is separated and becomes cloud and mist, while the clearest and lightest part is left, sweetened by the sun heating and boiling it. Everything is sweetened by boiling. So long as it is scattered and does not mass together, it remains floating in the air. But when it is gathered and collected suddenly by the assault of contrary winds, then it falls wherever there happens to be the densest cloud. This is most likely to happen when a wind has gathered some clouds together and is driving them along and then another wind suddenly confronts it with another mass of clouds. Then the first cloud is stopped and the

following ones pile up on it till it becomes thick and black and dense, and its weight causes it to turn to rain and fall. Rain water, therefore, is likely to be the best of all water, but it needs to be boiled and purified. If not, it has a foul smell and causes hoarseness and deepness of the voice in those that drink it.

Water from snow and ice is always harmful because, once it has been frozen, it never regains its previous quality. The light, sweet and sparkling part of it is separated and vanishes leaving only the muddiest and heaviest part. You may prove this, if you wish, by measuring some water into a jar and then leaving it out in the open air on a winter's night in the coldest spot you can find. Next morning bring it back into the warmth again and, when it has thawed, measure it a second time. You will find the quantity considerably less. This shows that in the process of freezing, the lightest and finest part has been dried up and lost, for the heaviest and densest part could not disappear thus. For this reason I consider such water to be the most harmful for all purposes.

9. The effect of drinking water collected from many different sources, that is, from large rivers fed by smaller streams and from lakes into which many streams flow from different directions, is to cause a propensity to stone, gravel in the kidneys, strangury, pain in the loins and rupture. The same is true of water brought long distances from its source. The reason for this is that no two sorts of water can be alike but some will be sweet, some salt and astringent and some from warm springs. When they are all mixed they quarrel with one another and the strongest is always the dominant. But each one has not always the same strength and sometimes one is dominant, sometimes another according to which wind is blowing. One will be made strong by the north wind, another by the south and so on. Such water will leave a sediment of sand and slime at the bottom of the jar and it is by drinking this that the diseases mentioned above are caused. There are, however, certain exceptions and these I shall detail.

Those whose stomachs are healthy and regular, and whose bladders are not subject to inflammation, nor in whom the neck of the bladder is overmuch obstructed, pass water easily and nothing collects in the bladder. But if the belly is liable to fever the same must be true of the bladder, and when this organ is heated with fever, the neck of the bladder becomes inflamed and does not allow the urine to pass which instead becomes heated and condensed. The finest and clearest part is separated, passes through and is voided. The densest and cloudiest part is gathered together and precipitates in small pieces at first and then in larger ones. The gravel formed is rolled round by the urine and coalesces to form a stone. When water is passed this falls over the neck of the bladder, and being pressed down by the pressure of the urine, prevents the urine from being passed. Great pain is thus caused. As a result, children suffering from stone rub or pull at their private parts because they think that in them lies the cause why they cannot make water. The fact that people who suffer from stone have very clear urine is proof that the densest and muddiest part remains in the bladder and collects there. This is the explanation of most cases of this disease but, in children, stones may also be caused by milk. If milk is not healthy but too warm and bilious-looking, it heats the stomach and the bladder and the urine is heated and a similar result is produced to that already described. Indeed, I assert that it is better to give children wine watered down as much as possible for this neither burns the veins nor dries them up too much. Female children are less liable to stone because the urethra is short and wide and the urine is passed easily. Neither do they masturbate as the males do, nor touch the urethra. In the female the urethra is short; in males it is not straight and it is narrow as well. Moreover, girls drink more than boys.

10. Now let us consider the seasons and the way we can predict whether it is going to be a healthy or an unhealthy year. It is most likely to be healthy if the signs observed at the rising and the setting of the stars occur normally, when there is rain in the autumn, when the winter is moderate being neither too mild nor excessively cold, and when rain falls seasonally in spring and in summer. But if the winter be dry with northerly winds prevailing and the spring wet with southerly

winds, the summer will of necessity be feverish and productive of ophthalmia and dysentery. For when stifling heat succeeds while the ground is still wet from the spring rains and southerly winds, the heat will be twice as great. Firstly because of the soaked warm earth and secondly because of the blazing sun; and, moreover, men's stomachs will not be toughened nor the brain firm. In such a spring the flesh cannot but become flabby and this predisposes to acute fevers, especially in those of phlegmatic constitution. Dysentery is likely to attack women and those of watery constitution. Should the etesian winds blow and there is bad weather and rain at the rising of the Dog Star, then it may be hoped that these bad conditions will come to an end and that the autumn will be a healthy one. But if there is no amelioration in the conditions there is a danger of fatalities among women and children; the elderly are in the least danger. Those who recover are liable to quartan fevers in which dropsy may supervene.

If the winter is wet and mild with southerly winds and this is followed by a wintry dry spring with the wind in the north, the effect will be as follows. First, women who happen to be pregnant and approaching term in the spring are likely to have miscarriages. Or, if they do give birth, the babies are so weak and sickly that either they die at once or, if they survive, they are frail and weak and very liable to disease. The men are liable to dysentery and dry ophthalmia, while some will suffer from catarrh of the head which may spread to the lungs. It is those who are full of phlegm, as well as the women, who are likely to suffer from dysentery since the phlegm flows down from the brain on account of their moist constitutions. On the other hand, those who are full of bile suffer from dry ophthalmia on account of the warmth and dryness of the flesh, while the old, owing to the permeability and exhaustion of the blood-vessels, suffer from catarrh. This last illness may prove suddenly fatal to some, while others are afflicted with a right- or left-sided hemiplegia. The explanation of these diseases is this. When the winter is warm with wet south winds neither the brain nor the blood-vessels become consolidated. Thus, when spring comes with dry cold northerly winds, the brain

becomes stiff and cold just when it ought to thaw and become
purified by running of the nose and hoarseness. It is the sudden
change when the heat of summer comes that is responsible for
these diseases.

Districts which are well situated with regard to the sun and
the winds and which have a good water supply are the least
affected by such changes in the weather; those badly situated
with regard to the sun and the winds and which draw their
water from marsh or lake, the most. If the summer be dry,
diseases are short lived, but if it is wet they last long and there
is the danger of a sore appearing on the slightest pretext if the
skin is broken. Diarrhoea and dropsy occur towards the
termination of illnesses under such conditions because the
bowels do not dry.

If the summer is rainy with southerly winds and the autumn
similar, the winter will necessarily be unhealthy. Those of
phlegmatic constitution and those over forty years old may
suffer from *causus*, while those who are full of bile suffer from
pleurisy and pneumonia. If the summer is dry with northerly
winds and the autumn wet with the wind in the south, the
winter brings a danger of headache and gangrene of the brain.
Further, there is likely to be hoarseness, running at the nose
and cough and, in some cases, consumption. If the autumn is
rainless with northerly winds and there is rain neither under
the Dog Star nor at Arcturus, this weather suits best those who
are naturally phlegmatic and of a watery constitution and also
women. But it is most inimical to those of a bilious disposition
because they become dried up too much. This produces dry
ophthalmia and sharp fevers which last a long time and also, in
some cases, 'black bile' or melancholy. The reason for this is
found in the drying up of the more fluid part of the bile while
the denser and more bitter part is left behind. The same is
true of the blood. But these changes are beneficial to those of
phlegmatic habit so that they become dried up and start the
winter braced up instead of relaxed.

11. Anyone making observations and drawing deductions
on these lines can foretell most of the effects which follow
changes in the weather. It is particularly necessary to take

precautions against great changes and it is inadvisable to give a purge, to cauterize or to cut any part of the belly until at least ten days have passed after such a change. The most dangerous times are the two solstices, especially mid-summer, and the equinoxes. Both of these latter times are considered dangerous but more especially the autumnal one. Care must also be taken at the rising of certain stars, particularly the Dog Star and Arcturus. Similarly, discretion must be exercised at the setting of the Pleiads. It is at such times that the crisis is reached in the course of diseases; some prove fatal and some are cured, but all show some kind of change and enter a new phase.

12. I now want to show how different in all respects are Asia and Europe, and why races are dissimilar, showing individual physical characteristics. It would take too long to discuss this subject in its entirety but I will take what seem to me to be the most important points of difference.

Asia differs very much from Europe in the nature of everything that grows there, vegetable or human. Everything grows much bigger and finer in Asia, and the nature of the land is tamer, while the character of the inhabitants is milder and less passionate. The reason for this is the equable blending of the climate, for it lies in the midst of the sunrise facing the dawn. It is thus removed from extremes of heat and cold. Luxuriance and ease of cultivation are to be found most often when there are no violent extremes, but when a temperate climate prevails. All parts of Asia are not alike, but that which is centrally placed between the hot and the cold parts is the most fertile and well wooded; it has the best weather and the best water, both rain water and water from springs. It is not too much burnt up by the heat nor desiccated by parching drought; it is neither racked by cold nor drenched by frequent rains from the south or by snow. Crops are likely to be large, both those which are from seed and those which the earth produces of her own accord. But as the fruits of the latter are eaten by man, they have cultivated them by transplanting. The cattle raised there are most likely to do well, being most prolific and best at rearing their young. Likewise, the men are well made, large and with good physique. They differ little among themselves in size and physical development. Such a land resembles the spring time in its character and the mildness of the climate.

*

16.* So much for the differences of constitution between the inhabitants of Asia and of Europe. The small variations of climate to which the Asiatics are subject, extremes both of heat and cold being avoided, account for their mental flabbiness and cowardice as well. They are less warlike than Europeans and tamer of spirit, for they are not subject to those physical changes and the mental stimulation which sharpen tempers and induce recklessness and hot-headedness. Instead they live under unvarying conditions. Where there are always changes, men's minds are roused so that they cannot stagnate. Such things appear to me to be the cause of the feebleness of the Asiatic race, but a contributory cause lies in their customs; for the greater part is under monarchical rule. When men do not govern themselves and are not their own masters they do not worry so much about warlike exercises as about not appearing warlike, for they do not run the same risks. The subjects of a monarchy are compelled to fight and to suffer and die for their masters, far from their wives, their children and friends. Deeds of prowess and valour redound to the advantage and advancement of their masters, while their own reward is danger and death. Moreover, such men lose their high-spiritedness through unfamiliarity with war and through sloth, so that even if a man be born brave and of stout heart, his character is ruined by this form of government. A good proof of this is that the most warlike men in Asia, whether Greeks or barbarians, are those who are not subject races but rule themselves and labour on their own behalf. Running risks only for themselves, they reap for themselves the rewards of bravery or the penalties of cowardice. You will also find that the Asiatics differ greatly among themselves, some being better

* At this point some paragraphs have been lost, and the order of what remains is uncertain.

and some worse. This follows from the variations of climate to which they are subject, as I explained before.

*

13. Such then is my opinion of Egypt and Libya. I will now discuss the area to the east-north-east as far as Lake Maeotis,* for this is the boundary between Europe and Asia. The people inhabiting these regions differ more among themselves than those discussed previously on account of the changeability of the weather and the nature of the terrain. And what is true of the soil is true of the men. Where the weather shows the greatest and the most frequent variations, there the land is wildest and most uneven. You will find mountains, forests, plains and meadows. But where there is not much difference in the weather throughout the year, the ground will be all very level. Reflection will show that this is true of the inhabitants too. Some men's characters resemble well-wooded and watered mountains, others a thin and waterless soil, others plains or dry bare earth. Climates differ and cause differences in character; the greater the variations in climate, so much the greater will be differences in character.

14. I will leave out the minor distinctions of the various races and confine myself to the major differences in character and custom which obtain among them. First the Macrocephali; no other race has heads like theirs. The chief cause of the length of their heads was at first found to be in their customs, but nowadays nature collaborates with tradition and they consider those with the longest heads the most nobly born. The custom was to mould the head of the newly-born children with their hands and to force it to increase in length by the application of bandages and other devices which destroy the spherical shape of the head and produce elongation instead. The characteristic was thus acquired at first by artificial means, but, as time passed, it became an inherited characteristic and the practice was no longer necessary. The seed comes from all parts of the body, healthy from the healthy parts and sickly from the sickly. If therefore bald parents usually have bald

*The Sea of Azov.

children, grey-eyed parents grey-eyed children, if squinting parents have squinting children, why should not long-headed parents have long-headed children? But in fact this does not happen as often as before, because the custom of binding the head has also become obsolete through intercourse with other peoples.

15. I pass now to consider the people who live near the river Phasis.* Their land is marshy, warm, wet and thickly covered with vegetation. Violent rainstorms occur there frequently at all seasons of the year and the inhabitants live in the marshes. Their houses are built on the water of wood and reeds and they do very little walking to go to town or to market, but sail up and down along the many canals in dug-out canoes. They drink warm stagnant water which has been rotted by the sun and swollen by the rains, and the Phasis itself is the most sluggish and stagnant of all rivers. The crops that grow there are all poor, feeble and do not ripen well owing to the super-abundance of water which interferes with the ripening process. The ground is often covered with mist. As a result of this the Phasians have peculiar constitutions. They are big and stout and their joints and veins are obscured by flesh. Their skin is yellowish as if they had jaundice and their voices, because they breathe the air which is moist and damp and not clean, are the deepest known. They have little stamina but become quickly tired. The climate varies very little and the prevailing winds are southerly, except for one local breeze which some-times blows a stiff warm gale. They call this wind the Kenkhron. The north wind never blows hard even when it does blow.

17.† In Europe, on the other hand, and living round Lake Maeotis, there is a special race of Scythians which differs from all other peoples. They go by the name of Sauromatae. Their women ride horses and shoot arrows and hurl javelins from horseback and they fight in campaigns as long as they remain virgins. Nor do they lose their virginity until they have killed three of their enemies and have offered such sacrifices as are prescribed by ritual law. But once a woman has taken to her-self a husband she does not ride again unless military necessity

*Rion.   †See note on p. 160.

should require their total forces to take to the field. The women have no right breast since their mothers heat a specially made iron and apply it to the breast while they are still children. This prevents the breast from growing and all the strength and size of it go into the right arm and shoulder instead.

18. As regards the appearances of other tribes of Scythians, the same is true of them as is true of the Egyptians, namely, that they have certain racial characteristics, but differ little among themselves. They differ, however, from the Egyptians in that their peculiarities are due to cold instead of to heat. The so-called Scythian desert is a grassy plain devoid of trees and moderately watered, for there are large rivers there which drain the water from the plains. Here live the Scythians who are called nomads because they do not live in houses but in wagons. The lighter wagons have four wheels but some have six, and they are fenced about with felt. They are built like houses, some with two divisions and some with three, and they are proof against rain, snow and wind. The wagons are drawn by two or three yokes of hornless oxen; hornless because of the cold. The women live in these wagons while the men ride on horseback, and they are followed by what herds they have, oxen and horses. They stay in the same place as long as there is enough grass for the animals but as soon as it fails they move to fresh ground. They eat boiled meat and drink the milk of mares, from which they also make a cheese.

19. So much then for their mode of life and customs. As regards their physical peculiarities and the climate of their lands, the Scythian race is as far removed from the rest of mankind as can be imagined and, like the Egyptians, they are all similar to one another. They are the least prolific of all peoples and the country contains very few wild animals and what there are are very small. The reason for this is their situation in the far north under the Rhipaean mountains from which the north wind blows. The sun shines most brightly towards its setting in the summer and then it warms them only for a very short time and not very much. In addition, the winds from warm lands do not reach as far, as a rule, or, if they do,

they are weak. Instead, northerly winds, chilled with snow and ice and charged with great rains, blow continuously and never leave the mountains which makes them most inhospitable. During the daytime mist often covers the plains where the people live and, in fact, winter is nearly continuous all the year round. The summer lasts only a few days and these are not very summery for the plains are highly situated, bare of trees and are not engirdled by mountains, but slope from the north. The only wild animals found there are those small enough to shelter underground. The cold weather together with the barrenness of the ground, which affords neither warmth nor shelter, prevents their growth. There are no great nor violent changes with the seasons, the climate remaining very much the same all the year round. The people differ little in physique as they always eat similar food, wear the same clothes winter and summer, breathe moist thick air, drink water from snow and ice and do no hard work. The body cannot become hardened where there are such small variations in climate; the mind, too, becomes sluggish. For these reasons their bodies are heavy and fleshy, their joints are covered, they are watery and relaxed. The cavities of their bodies are extremely moist, especially the belly, since, in a country of such a nature and under such climatic conditions, the bowels cannot be dry. All the men are fat and hairless and likewise all the women, and the two sexes resemble one another. Owing to the lack of variation in the weather, the coagulation of the seed is not prevented or impeded unless there is some violent injury or inter-current disease.

20. As a proof of this moistness of the constitution, I may instance the following. You will find that the majority of the Scythians, especially those who are nomads, are cauterized on the shoulders, arms, wrists, chests, hips and loins. This is done simply for the softness and moistness of their constitutions because otherwise they could neither bend their bows nor put any weight into throwing the javelin. But when they have been cauterized the moisture is dried out of their joints and their bodies become more sinewy and stronger and their joints may then be seen. They grow up flabby and stout for two

reasons. First because they are not wrapped in swaddling clothes, as in Egypt, nor are they accustomed to horse-riding as children which makes for a good figure. Secondly, they sit about too much. The male children, until they are old enough to ride, spend most of their time sitting in the wagons and they walk very little since they are so often changing their place of residence. The girls get amazingly flabby and podgy. The Scythians have ruddy complexions on account of the cold, for the sun does not burn fiercely there. But the cold causes their fair skins to be burnt and reddened.

21. People of such constitution cannot be prolific. The men lack sexual desire because of the moistness of their constitution and the softness and coldness of their bellies, a condition which least inclines men to intercourse. Moreover, being perpetually worn out with riding they are weak in the sexual act when they do have intercourse. These reasons suffice as far as the men are concerned. In the case of the women, fatness and flabbiness are also to blame. The womb is unable to receive the semen and they menstruate infrequently and little. The opening of the womb is sealed by fat and does not permit insemination. The women, being fat, are easily tired and their bellies are cold and soft. Under such conditions it is impossible for the Scythians to be a prolific race. As a good proof of the sort of physical characteristics which are favourable to conception, consider the case of serving wenches. No sooner do they have intercourse with a man than they become pregnant, on account of their sturdy physique and their leanness of flesh.

22. Further, the rich Scythians become impotent and perform women's tasks on an equal footing with them and talk in the same way. Such men they call Anarieis. The Scythians themselves attribute this to a divine visitation and hold such men in awe and reverence, because they fear for themselves. Indeed, I myself hold that this and all other diseases are equally of divine origin and none more divine nor more earthly than another. Each disease has a natural cause and nothing happens without a natural cause. My own explanation of this disability of the Scythians is this. As a result of horse-riding they are afflicted with varicosity of the veins because

their feet are always hanging down from their mounts. This is followed by lameness and, in severe cases, those affected drag their hips. They treat themselves by their own remedy which is to cut the vein which runs behind each ear. The haemorrhage which follows causes weakness and sleep and after this some, but not all, awake cured. My own opinion is that such treatment is destructive of the semen owing to the existence of vessels behind the ears which, if cut, cause impotence and it seems to me that these are the vessels they divide. Consequently when they come into the presence of their wives and find themselves impotent, they do not perhaps worry about it at first, but when after the second and third and more attempts the same thing happens, they conclude that they have sinned against the divinity whom they hold responsible for these things. They then accept their unmanliness and dress as women, act as women and join with women in their toil.

That it is the rich Scythians, those of the noblest blood and the greatest wealth, and not their inferiors, who suffer from this disease is due to horse-riding. The poor suffer less because they do not ride. Yet, surely, if this disease is more to be considered a divine visitation than any other, it ought to affect not only the rich but everyone equally. Rather, the poor should be specially liable to it if the gods really do delight in honours and the admiration of men and bestow favours in return. It is the rich who make frequent sacrifice and dedication to the gods because they have the means. The poor, being less well provided with goods, sacrifice less and accompany their prayers with complaint. Surely it is the poor and not the rich who should be punished for such sins. Really, of course, this disease is no more of 'divine' origin than any other. All diseases have a natural origin and this peculiar malady of the Scythians is no exception. The same thing happens in other races. Those who ride the most suffer most from varicose veins, pain in the hips and gout and they are the less able to perform their sexual functions. This is the fate of the Scythians. They are the most effeminate race of all mankind for the reasons I have given, and because they always wear trousers and spend so much of their time on horseback so that they do not handle their private parts, and, through cold and exhaustion, never have even the desire for sexual intercourse. Thus they have no sexual impulses in the period before they lose their virility.

23. The remaining peoples of Europe differ widely among themselves both in size and appearance owing to the great and frequent climatic changes to which they are subject. Hot summers and hard winters, heavy rains followed by long periods of drought, all these occasion variations of every kind. It is reasonable that these changes should affect reproduction by variations in the coagulability of the semen so that its nature is different in summer and winter, in rainy weather and times of drought. I believe this to be the reason for the greater variation among individuals of the European races, even among the inhabitants of a single city, than is seen among Asiatics and also why they vary so much in size. When the weather changes often, abnormalities in the coagulation of the semen are more frequent than when the weather is constant. A variable climate produces a nature which is coupled with a fierce, hot-headed and discordant temperament, for frequent fears cause a fierce attitude of mind whereas quietness and calm dull the wits. Indeed, this is the reason why the inhabitants of Europe are more courageous than those of Asia. Conditions which change little lead to easy-going ways; variations to distress of body and mind. Calm and an easy-going way of living increase cowardice; distress and pain increase courage. That is one reason for the more warlike nature of Europeans. But another cause lies in their customs. They are not subjects of a monarchy as the Asiatics are and, as I have said before, men who are ruled by princes are the most cowardly. Their souls are enslaved and they are unwilling to risk their own lives for another's aggrandisement. On the other hand, those who govern themselves will willingly take risks because they do it for themselves. They are eager and willing to face even the worst of fates when theirs are the rewards of victory. It is clear, then, that the tradition of rule has no small influence on the courage of a people.

24. In general it may be said that these are the differences

between Europe and Asia. There exist in Europe, then, people differing among themselves in size, appearance and courage, and the factors controlling those differences are those I have described. Let me summarize this plainly. When a race lives in a rough mountainous country, at a high elevation, and well watered, where great differences of climate accompany the various seasons, there the people will be of large physique, well-accustomed to hardihood and bravery, and with no small degree of fierceness and wildness in their character. On the other hand, in low-lying, stifling lands, full of meadows, getting a larger share of warm than cold winds, and where the water is warm, the people will be neither large nor slight, but rather broad in build, fleshy and black-haired. Their complexions are dark rather than fair and they are phlegmatic rather than bilious. Bravery and hardihood are not an integral part of their natural characters although these traits can be created by training. The people of a country where rivers drain the surface water and rain water have clear complexions and good health. But where there are no rivers and the drinking water is taken from lakes or marshes, the people will necessarily be more pot-bellied and splenetic. People who live in countries which are high, level, windswept and rainy tend to be of large stature and to show little variation among themselves. They are also of a less courageous and less wild disposition. In countries where there is a light waterless soil devoid of trees and where the seasons occasion but small changes in climate, the people usually have hard sinewy bodies, they are fair rather than dark and they are strong-willed and headstrong in temperament. Places where changes of weather are most frequent and of the greatest degree show the greatest individual differences in physique, temperament and disposition among the inhabitants.

The chief controlling factors, then, are the variability of the weather, the type of country and the sort of water which is drunk. You will find, as a general rule, that the constitutions and the habits of a people follow the nature of the land where they live. Where the soil is rich, soft and well-watered and where surface water is drunk, which is warm in summer and cold in winter, and where the seasons are favourable, you will find the people fleshy, their joints obscured, and they have watery constitutions. Such people are incapable of great effort. In addition, such a people are, for the most part, cowards. They are easy-going and sleepy, clumsy craftsmen and never keen or delicate. But if the land is bare, waterless and rough, swept by the winter gales and burnt by the summer sun, you will find there a people hard and spare, their joints showing, sinewy and hairy. They are by nature keen and fond of work, they are wakeful, headstrong and self-willed and inclined to fierceness rather than tame. They are keener at their crafts, more intelligent and better warriors. Other living things in such a land show a similar nature. These, then, are the most radically opposed types of character and physique. If you draw your deductions according to these principles, you will not go wrong.

PROGNOSIS

The importance of being able to foretell the course of an illness, and an account of the significance of various signs.

1. It seems to be highly desirable that a physician should pay much attention to prognosis. If he is able to tell his patients when he visits them not only about their past and present symptoms, but also to tell them what is going to happen, as well as to fill in the details they have omitted, he will increase his reputation as a medical practitioner and people will have no qualms in putting themselves under his care. Moreover, he will the better be able to effect a cure if he can foretell, from the present symptoms, the future course of the disease.

It is impossible to cure all patients; that would be an achievement surpassing in difficulty even the forecasting of future developments. But seeing that men die before the physician is able to bring his skill to grapple with the case – some owing to the violence of the disease die before they have summoned the doctor, some as soon as he arrives; some live one day, others a little longer – in view of this, an understanding of such diseases is needed. One must know to what extent they exceed the strength of the body and one must have a thorough acquaintance with their future course. In this way one may become a good physician and justly win high fame. In the case of patients who were going to survive, he would be able to safeguard them the better from complications by having a longer time to take precautions. By realizing and announcing beforehand which patients were going to die, he would absolve himself from any blame.

2. The signs to watch for in acute diseases are as follows. First study the patient's *facies*; whether it has a healthy look and in particular whether it be exactly as it normally is. If the patient's normal appearance is preserved, this is best; just as the more abnormal it is, the worse it is. The latter appear-

ance may be described thus: the nose sharp, the eyes sunken, the temples fallen in, the ears cold and drawn in and their lobes distorted, the skin of the face hard, stretched and dry, and the colour of the face pale or dusky. Now if at the beginning of an illness the face be such and one's judgement lacks confirmation from other signs, the patient should be asked whether he has suffered from insomnia, from severe diarrhoea, or if he has ravening hunger. If he admits to any of these things, the case must be judged less severe than if it were otherwise, for where the facial appearance is due to any of these causes a crisis will be reached in a day and a night. But if he admits none of these things, and if there is no improvement within the prescribed time, it must be realized that this sign portends death.

Should the illness have passed the third day before the face assumes this appearance, the same questions as I mentioned before should be asked, and an examination of the whole body made for other signs, paying particular attention to the eyes. For if they avoid the glare of light, or weep involuntarily, or squint, or the one becomes smaller than the other, or if the whites are red or livid or show the presence of tiny dark veins, or if bleariness appear around the eyes, or if the eyes wander, or project, or are deeply sunken, or if the whole complexion of the face be altered; then all these things must be considered bad signs and indicative of death.

The appearance of the eyes in sleep should also be noted, for if some of the white shows when the eyes are closed, so long as it is not due to diarrhoea, the taking of drugs, or the normal habit in sleep, it is a bad sign and especially fatal. If the eyelid becomes swollen or livid, or likewise the lip or the nose, together with one of the other signs, it may be known that death is at hand. It is also a fatal sign if the lips are parted and hang loose and become cold and white.

3. When the physician visits the patient, he should find him lying on one side or the other, with his hands, neck and legs slightly bent, and with the whole body lying relaxed. For this is how most healthy people lie. The best manner of lying in bed is that which most nearly resembles the manner of healthy

people. It is not so good if the patient lies on his back with his hands and legs extended; while if he should have fallen forwards away from the bed towards his feet, that is worse still.

If he should be found with his feet uncovered, unless they are exceptionally warm, and with his hands and legs flung about at random, it is a bad sign because it is evidence of restlessness.

It is a fatal sign to sleep with the mouth continuously wide open, and if the patient lies on his back with his legs very much bent and intertwined. It is also bad if a patient sleeps on his stomach unless this is his normal habit when well; such a posture indicates delirium or abdominal pain.

For the patient to want to sit up when the disease is at its height is a bad sign in all acute diseases, but worst of all in cases of pneumonia. For a patient with fever to grind his teeth, unless this be a habit continued from childhood, is a sign of madness and death. If this occurs during delirium, it is a sign that the disease has already taken a fatal turn.

Inquiries should be made about any sore which has been discovered to ascertain whether it existed prior to the illness, or whether it has developed during the course of the illness. For, if the patient is about to die, before death the sore will become either livid and dry or pale and hard.

4. The following points about the gestures of the hands should be noted. In cases of acute fever or of pneumonia and in brain-fever and headache, it is a bad sign and portends death if any of the following things are noted: if the hands are waved in front of the face, or make grabs at the air, or pull the nap off cloth, or pull off bits of wool, or tear pieces of straw out of the wall.

5. Rapid breathing indicates either distress or inflammation in the organs above the diaphragm. Deep breaths taken at long intervals are a sign of delirium. If the expired air from the mouth and nostrils is cold, death is close at hand. Regular respiration is to be considered a most important indication of recovery in all the acute diseases which are accompanied by fever and reach the crisis within forty days.

6. Fits of sweating are excellent in all acute diseases when they occur on the critical days, and mark the final end of the

fever. They are also good when the whole body is involved, and show that the patient is taking the disease more easily. But those which conform to neither of these circumstances are of no advantage. The worst kinds of sweating are those which are cold and occur only round the head and neck; these, if accompanied by a high fever, mean death; if by a milder fever, a long illness.

7. The most satisfactory condition of the hypochondrium is when it is painless, but is soft and smooth on both sides. On the other hand, precautions must be taken if it is inflamed and painful, or taut, or if there is a difference in level between the two sides. Should there also be a throbbing in the hypochondrium, it is a sign of violent disturbance or delirium. In such cases, the appearance of the eyes should be noted; if the eyes move rapidly, it is highly probable that the patient is mad.

A hard and painful swelling of the hypochondrium which involves the whole of that area is a very bad sign. If it be only on one side, it is less fraught with danger if it is on the left side. When such swellings are present at the beginning of an illness, it is an indication of the danger of a speedy death; but if the patient lasts more than twenty days while the fever continues and without the swelling subsiding, then it will suppurate. In these cases, a violent epistaxis occurs during the first period and this is very helpful, but the patient should be asked whether he has a headache, or if his sight is dim. If either of these symptoms were present, it would incline to provoke the epistaxis. Epistaxis is more likely to occur in patients under the age of thirty-five.

Swellings which are soft, painless and pit on pressure with the finger, cause delayed crises but are less to be feared than the former kind. But if the fever continue for more than sixty days without the swelling subsiding, then an empyema is being formed. The same is also true of swellings in the belly.

In brief, then, painful hard large swellings mean danger of a speedy death; soft, painless swellings which pit on pressure mean protracted illness.

Swellings in the belly are less likely to be productive of abscess than those in the hypochondria, and those below the navel are the least likely to suppurate. Epistaxis is particularly to be expected in association with swellings in the upper parts. Whenever a swelling lasts a long time, the formation of an empyema at that site must be expected.

When suppuration occurs, the following points should be noted. Of those which point externally, the best sort are those which are small, bulge outwards as much as possible and come to a sharp head. The worst sort are those which are large and flat, and which do not come to anything like a sharp head. Of those which burst internally, the best are those which have no connection with the exterior, and which are localized, painless and show a uniform colouring all over the external surface. The best sort of pus is that which is white, smooth, homogeneous and least foul-smelling. That of the opposite sort is the worst.

8. All cases of dropsy arising from acute diseases are bad. For, besides not getting rid of the fever, they are particularly painful and liable to cause death. In most cases dropsy starts from the flanks and the loins, but sometimes from the liver. In those cases where the dropsy starts from the flanks and loins, the feet swell and long-lasting diarrhoea occurs which neither puts an end to the pain in the flanks and loins nor empties the belly. Where it arises from the liver, they have a desire to cough but produce no sputum worth mentioning, their feet swell, they pass nothing from the bowels but hard and painful stools passed with an effort, and swellings appear around the belly which come and go, sometimes on the right and sometimes on the left.

9. It is a bad sign if the head, hands and feet are cold while the belly and sides are warm. It is best that the body should be warm all over and equally soft.

The patient should turn over easily and be light when lifted up. If he should appear rather heavy, in the hands and feet as well as in the rest of the body, there is greater danger. If, in addition to this heaviness, the nails and fingers become livid, death is immediately to be expected. But if the fingers or the feet become completely black, this is less fatal than if they are livid. Nevertheless, other signs should be taken into consideration as well, for if the patient appears to be bearing up well under the disease, or if he displays in addition to this any of the signs which betoken recovery, it is probable that the disease will result in abscess formation with survival of the patient, although those parts will be lost which have turned black.

Drawing up of the testicles or private parts indicates distress or death.

10. As regards sleep, the patient should follow our natural habit and spend the day awake and the night asleep. If this habit be disturbed, it is not so good. Nevertheless, it is better that he should sleep during the morning and early afternoon than later. It is worst of all when he sleeps neither night nor day; it may be that pain and distress is keeping him awake, or this sign of insomnia may precede delirium.

11. It is best when the stools are soft and formed, and passed at the hour customary to the patient when in health; their bulk should be proportionate to the amount of food taken. Such stools indicate a healthy condition of the lower bowel. But if the stools be fluid, it is best that they should not be accompanied by a noise, nor passed in small quantities at frequent intervals; the continual getting up is exhausting for the patient and prevents him from sleeping. If he should pass large stools frequently, there is a risk of his fainting. He should, according to how much he eats, pass stools two or three times during the day and once during the night. The larger stool should be passed in the morning as he was accustomed. The stools should become more solid towards the crisis when the disease is being cured. They should be light brown and not too foul-smelling. It sometimes happens that round worms are passed with the stools toward the crisis when the disease is being cured.

In every illness, the belly should be loose and the stools of good size. It is a bad sign if the stools are very watery, or white, or particularly yellowish or frothy. It is also bad if they are small, sticky, white, yellowish and smooth. Signs more indicative of death are when they are dark, or livid, or oily, or rust-coloured and foul-smelling. Variety in the stools denotes a longer illness, but is no less a sign of a fatal outcome. Such stools are those which are full of shreds, bloody, bilious, green and dark stools; sometimes such constituents are passed together, sometimes separately.

It is best to emit wind without a noise or breaking wind; but it is better to emit it even with a noise than to repress or smother it. All the same, wind emitted in this manner indicates that there is something wrong internally, or that the patient is delirious, if, at least, the emission of wind is involuntary.

Pains and swellings in the hypochondria, if they are fresh and not accompanied by inflammation, are dispersed by a rumbling gathering of wind in the hypochondrium, especially if it be passed through the body and voided with the stools and urine. It may be passed through by itself. It is also a good thing if the gathering of wind moves down to the lower regions.

12. Urine is best when there is a white, smooth, even deposit in it the whole time up to the crisis of the disease, for this indicates recovery and a short illness. If there should be intervals when clear urine is passed, and the white, smooth, even deposit appears only at times, this means that the illness will be prolonged and that recovery is less certain. If the urine should be pink with a pink smooth sediment, although such indicates an illness even longer than in the previous case, it is a certain sign of recovery. Sediment like barley-meal in the urine is bad, and it is even worse if the sediment resembles flakes. Thin white sediment is a very bad sign, and it is even worse if it resembles bran. Clouds suspended in the urine constitute a good sign if they are white, a bad one if they are dark.

So long as the urine is thin and yellowish-red, the disease is not ripened. If the illness is prolonged and the urine remain of that colour, there is a danger that the patient may not last out till ripening occurs. Urine which is foul-smelling or watery or dark or thick is more a sign of death. In the case of men and women, dark urine is worst; in the case of children, watery urine.

When a patient continues to pass thin raw urine for a long time and the other signs indicate recovery, the formation of an abscess should be expected in the parts below the diaphragm.

When grease forms patterns like cobwebs on the surface of the urine, this constitutes a warning, for it is a sign of wasting. When urine contains clouds, it should be noted whether they are towards the top or the bottom, and what is their colour. Those which sink and have the colours previously mentioned as favourable are to be judged a good sign. Those which rise and have the colour said to be unfavourable constitute a bad sign. You must not be deceived if these appearances result merely from a diseased condition of the bladder, for they may then indicate not a disease of the whole body, but merely of that organ.

13. The most helpful kind of vomiting is that in which the matter brought up consists of phlegm and bile, as well-mixed as possible, and is neither thick nor particularly great in quantity. If it is not well-mixed, it is less good. The vomiting of dark green, livid or dark material, no matter which of these colours, must be considered a bad sign. If the same patient should vomit material of all these colours, his condition is already fatal. The quickest death is denoted by the vomiting of livid matter if it has a foul smell. All rotten and foul odours coming from vomited material are bad.

14. In all diseases which affect the lungs and sides, sputum should be brought up early and, in appearance, the yellow matter should be thoroughly mixed with the sputum. It is not so good if it only comes about some while after the beginning of the pain, that the sputum is brought up and it is yellow, or light brown, or the cause of much coughing, or if it be not thoroughly mixed. It is a sign of danger if the yellow matter is not diluted; and white, sticky and nummular sputum is not beneficial. It is worse if it should be a marked pale green and frothy. If it should be so undiluted as to appear dark, this is even worse still. It is also bad if the lungs are not cleared and nothing is produced, but the throat remains full of bubbling matter.

In all diseases of the lungs, running at the nose and sneezing is bad, whether it existed before the illness or supervened during its course. But in other diseases which are likely to prove fatal, sneezing is beneficial. In cases of pneumonia, the production at the beginning of the illness of yellow sputum mixed with a little blood is a good indication of recovery. But when this occurs on or after the seventh day, it is less certainly good. All sputa are bad which do not relieve the pain; the worst are those which are dark in colour as stated above. The production of any sputum which relieves pain is rather better.

15. When aches arising in these regions are not relieved by the production of sputum or evacuation of the bowels, or by venesection or the administration of drugs and special regimens, you must know that an empyema is present. Those empyemas which begin to suppurate while the sputum is still bilious are especially signs of a fatal issue, whether the bilious matter is brought up separately from the pus or together with it.

Most specially, if the empyema appears to start from sputum of this sort when the disease is in its seventh day, the patient who brings up such sputum may be expected to die on the fourteenth day, unless a good sign makes its appearance. The good signs which may appear are these: to bear the illness easily, to have good respiration, to be free from pain, to cough up sputum easily, to have the body evenly warm and soft all over, not to suffer from thirst, to have the urine, stools, sleep and sweating of the types described above as good. If all these signs appear the patient is not likely to die; but if only some of these signs appear, he may die although he will live longer than fourteen days. The opposites of these are bad signs: to bear the illness hardly, to draw deep and frequent breaths, to suffer continued pain, to have difficulty in coughing up sputum, to have violent thirst, to have the body unevenly warm with the abdomen and sides very warm and the forehead, hands and feet cold, to have the urine, stools, sleep and sweating of the types described above as bad. If any of these signs appear subsequent to the bringing up of sputum of this description, the patient may die in less than fourteen days, either on the ninth or eleventh. This is the inference which should be drawn from observing this kind of sputum; it is particularly likely to indicate death and may not give the patient his fourteen days.

The most reliable forecast is that which takes into account the good and the bad signs which appear in addition. Other empyemata burst as a rule, some on the twentieth day, some on the fortieth and some reach sixty days.

16. The beginning of an empyema may be reckoned for calculation from the day on which the patient first had a fever, or when he had a rigor, or when he said that a heaviness replaced the pain in the spot where he feels discomfort. These things occur at the start of an empyema. The discharge of pus must be expected according to the stated intervals reckoned from this day.

In cases where the empyema is unilateral, the patient should be made to turn over on the side affected and then asked whether he has an ache in that side. Or, if one side be hotter than the other, he should be made to lie on the healthy side and then be asked if he feels as if a weight were hanging on him from above. If such is the case, the empyema is on which-ever side the heaviness is felt.

17. All empyemata may be recognized by the following signs. First of all, the fever does not intermit, but remitting a little during the day, becomes more acute at night. Many fits of sweating occur. A desire to cough is aroused, but nothing is brought up to speak of. The eyes become sunken, and the cheeks are flushed. The finger-nails become curved and the fingers become warm, especially at their tips. Swellings which come and go are observed in the feet. Blisters form on the body and the patients show no desire for food.

Chronic empyemata show these signs and considerable reliance may be placed in them. Those which will not last long are indicated by their showing the sort of signs which appear at the start of the empyema, and also by the patient suffering somewhat from difficulty with breathing. Whether it will burst sooner or later may be determined from the following signs – if pain occurs at the beginning, and dyspnoea, cough and expectoration continue, bursting may be expected in twenty days or even less. If the pain be less acute and the other signs are normal, bursting may be expected after twenty days. Pain, dyspnoea and expectoration must always precede the evacuation of pus.

Patients with empyema who are most likely to survive are those whom the fever leaves on the same day as the abscess bursts, and who quickly regain their appetite, lose their thirst and who pass small firm stools, and from whom white smooth pus, all of the same colour and unmixed with phlegm, flows out and is cleared away without pain or coughing. These are the best signs and patients who show them speedily recover, and failing these, the best signs are their nearest approximations.

Those patients die when the fever does not leave them on the same day as the abscess bursts, but in whom, after an apparent departure, it reappears and gives them a high fever. They suffer from thirst, lack of appetite and diarrhoea, and the pus is greenish-yellow and livid, or phlegmatic and frothy. When all these signs appear, death is certain; when some appear, but not all, the patient may die or he may recover after a prolonged illness. But indications should be drawn not only from the special signs concerned with the empyema itself, but also from all the other signs as well.

18. When abscess formation from pneumonic conditions occurs near the ears and suppurates downwards with the production of a fistula, the patient recovers.

Some complications must be suspected under the following conditions: if the fever is continuous and the pain incessant, if the sputum does not appear normal, and if the stools are neither bilious nor loose and homogeneous, if the urine is not thick and containing much sediment. In such cases, if the other signs of recovery are favourable, the formation of abscesses can be expected. These sometimes occur in the lower regions, in cases where some of the phlegm is located near the hypochondrium; sometimes in the upper regions where the hypochondrium remains soft and painless. In such cases, the patient, having for some time suffered from dyspnoea, regains normal respiration without any apparent cause.

Abscess formation in the legs is always beneficial in severe and critical cases of pneumonia, but most specially so when it follows a change in the nature of the sputum. For if the swelling and pain come on when the sputum has become purulent instead of yellow and when it is being expectorated, this constitutes the surest sign that the patient is going to recover and that the abscess will quickly become painless and resolve. But if the sputum is not expectorated well and urine with a satisfactory sediment does not appear, there is a danger that the limb may become lame or give a good deal of trouble.

If the abscess disappears without any expectoration of sputum while the fever continues, it is a bad sign for there is a danger that the patient may become delirious and die.

It is the older people who are more likely to die when empyemata complicate pneumonia, whereas in the case of empyema from other causes, death is more frequent among younger people.

19. When pain accompanied by fever attacks the loins and lower regions, they are specially fatal if the pain leaves the lower regions and fastens on the diaphragm. Attention should therefore be paid to the other signs. If another bad sign appears as well, the case is hopeless. But if other bad signs do not appear when the pain leaps up to the diaphragm, there is a good chance of an empyema forming.

It is always a bad sign if the bladder becomes hard and painful; most fatal if this is accompanied by continuous fever. The distress occasioned by the bladder alone is enough to kill the patient, while the bowels remain unopened under such circumstances except for the forcible passage of hard matter. The passage of urine resembling pus with a white smooth sediment terminates the condition. If there is no improvement in the urine and the bladder does not become soft, and if the fever is continuous, the patient is likely to die early in the disease. This condition occurs most frequently in children between the ages of seven and fifteen.

20. Fevers reach their crises in the same number of days whether the patient survives or dies. The mildest fevers, and those which give the surest indications of recovery, cease on

or before the fourth day. Those which are the most severe and accompanied by the worst signs cause death on the fourth day or earlier. The first bout of a fever ends in this period, the second lasts until the seventh day, the third till the eleventh day, the fourth till the fourteenth day, the fifth till the seventeenth day, the sixth till the twentieth day. In the case of most acute diseases, the bouts continue for twenty days, each bout adding four days at a time. But none of these periods can be computed in whole numbers exactly; neither the solar year nor the lunar month are of such a length as to be counted in whole numbers of days.

Subsequently, addition continues in the same way so that the first period contains thirty-four days, the second forty days and the third sixty days. It is very difficult to distinguish at the beginning between those fevers which are going to reach a crisis in a long period for they are very much alike in the way they start. However, the possibility should be borne in mind from the first day and reconsidered every time a period of four days is added and then the way in which the disease is developing will not escape you. Quartan fevers too follow the same pattern.

Those fevers which are going to reach the crisis in a short time are easier to recognize, for they show considerable differences from the start.

The patients who are going to recover have good respiration and no pain, they sleep at night and display other signs of recovery. Those who are going to die have dyspnoea, insomnia and delirium, and display other very bad signs. Thus, once this is recognized, calculations must be made which are based on the period and the appropriate additions as the disease moves towards its crisis. The crises which women undergo after childbirth follow the same plan too.

21. Severe continuous headache accompanied by fever is a certain sign of death if any of the other fatal signs occur as well. If there are no such signs, but the headache lasts more than twenty days while the fever continues, an epistaxis or some other abscession to the lower regions should be expected. An epistaxis or empyema may also be expected while the

headache is still young, especially if it is temporal or frontal. Epistaxis is more likely with patients under thirty-five; empyema with older men.

22. Acute earache accompanied by continuous and severe fever is a bad sign; there is a danger that the patient may become delirious and die. In view of the dangerous nature of this condition, special attention must be paid from the first day to any other signs. Younger men die on the seventh day or sooner from this malady; older men much more slowly for they are less liable to fever and delirium and for this reason their ears suppurate before they reach a fatal stage. Nevertheless, at such ages, relapse is usually fatal. The younger men die before the ear suppurates. When white pus flows from the ear, there is a chance that a young man may recover if some other good sign appears as well.

23. An ulcerated throat accompanied by fever is a bad sign and, if any other sign of those previously mentioned as bad appears as well, it may be said in advance that the patient is in danger. The worst kind of sore throat, and that which carries off those who suffer from them most quickly, is that which shows no obvious sign either in the throat or in the neck, but produces excessive pain and orthopnoea. Suffocation occurs on the first, second, third or fourth day. Those cases which are in other respects very similar and suffer pain, but in which the throat swells up and becomes inflamed, are also very fatal but the disease is more protracted than the previous sort.

When both the throat and the neck are inflamed, these sore throats last longer. Those suffering from them recover especially if a rash appears on the neck and chest and the erysipelas does not turn inwards. If the erysipelas does not disappear in the critical number of days, or if an external swelling does not appear, or if pus is not coughed up easily and without distress, this constitutes a sign of death or of a relapse of the inflammation. It is safest when the erysipelas turns outwards as much as possible; if it turns towards the lungs, it causes delirium and empyema usually follows.*

24. A relapse is to be expected in those cases where a fever

*A short passage which seems to be an interpolation is omitted here.

departs either without any sign of resolution appearing or if it departs on days other than the critical ones. Whenever a fever is prolonged while the patient appears likely to recover and suffers no pain by reason of any inflammation or any other apparent cause, an abscession accompanied by swelling and pain into one of the joints, most probably one of the lower ones, should be expected. Such abscessions occur particularly and in a shorter time in patients under the age of thirty years.

The formation of an abscession should be suspected at once if the fever lasts more than twenty days without remission. This is less likely to happen with older people in whom the fever lasts longer. If the fever is continued, such an abscession should be expected, but if it intermits and attacks in an irregular fashion and continues thus till autumn is at hand, it is likely to develop quartan periodicity. Just as people under thirty are specially liable to abscession, so those above that age are specially liable to quartan fevers.

It should be observed that during the winter abscessions are more likely and take longer to depart, but they are less liable to return.

When a patient with a fever which is unlikely to cause death professes a headache and a blackness before the eyes, or if heartburn accompany this, bilious vomiting will occur. If a rigor occurs as well, and the parts below the diaphragm are cold, vomiting will occur even sooner. If the patient takes any food or drink at this time, it will very quickly be brought up again. When such distress begins on the first day, the greatest distress will be on the fourth and fifth days; recovery will be about the seventh. In most cases, however, distress begins on the third day and the disease reaches its height on the fifth, departing on the ninth day or on the eleventh. When the distress begins on the fifth day, and in other respects the condition is similar to that previously described, the crisis occurs on the fourteenth day. These symptoms are specially common in tertian fevers in the case of adults. Younger people do suffer from them in tertians too, but more often in more continued fevers and genuine tertians.

When in this kind of fever the patient complains of headache, but instead of darkness before the eyes his sight becomes dim or is dazzled, and instead of heartburn there is contraction of the hypochondrium on one side or the other unaccompanied by either pain or inflammation, epistaxis is more likely to occur than vomiting. Even so, epistaxis is more probable in the young; those over thirty are less liable to it, but more liable to vomiting.

Children are likely to have convulsions if the fever is high and if they are constipated, if they are wakeful, frightened, cry and change colour, turning pale, livid or red. This most commonly happens in children under the age of seven. As they grow up and reach adult years, they are no longer likely to be attacked by convulsions in the course of a fever, unless one of the most severe and worst signs appears as well, as happens in inflammation of the brain. Whether the children and the others will recover or die must be judged by the whole total of signs as described in each case.

This concludes my remarks on acute diseases and those arising from them.

25. Anyone who is to make a correct forecast of a patient's recovery or death, or of the length of his illness, must be thoroughly acquainted with the signs and form his judgement by estimating their influence one on another, as has been described in speaking of urine, sputa and other subjects. The physician must be quick to think of the trend of any diseases that are epidemic from time to time, and the climatic conditions must not escape him. It should, however, be observed that the indications and signs have invariably the same force, the bad being always bad and the good good, in every year and under all climatic conditions. The truth of those described in this treatise has been proved in Libya, in Delos and in Scythia. It should therefore be realized that there is nothing remarkable in being right in the great majority of cases in the same district, provided the physician knows the signs and can draw the correct conclusions from them. There is no point in seeking the name of any disease which has not been mentioned, for all which reach their crisis in the periods described may be recognized by the same signs.

REGIMEN IN ACUTE DISEASES

The effect of various regimens upon an ailing body. Apparently a polemic treatise written to refute certain doctrines held by the neighbouring school of medicine at Cnidus.

1. The authors of the book called *Opinions from Cnidus* have given a correct account of the symptoms in patients suffering from various diseases and, in some cases, of the ultimate effects of the disease. Thus far indeed anyone might go, if he inquired diligently of each patient what his symptoms were, without being a physician. But these authors have omitted a great deal of what the physician should learn from his patient without his telling him; details which vary from case to case but the interpretation of which may sometimes be of vital importance.

2. Whenever their interpretation of the symptoms leads them to prescribe a cure, my opinions differ from theirs very considerably. Nor is this the only criticism I have to make, for, in addition, they employ too few remedies. Thus, apart from acute diseases, they generally prescribe opening medicine and recommend their patients whey and milk to drink.

3. Of course, if these remedies were satisfactory and were adapted to the diseases for which they were prescribed, I should think very highly of them, seeing that so few were sufficient. But this simply is not the case. Later writers, however, have approached the subject in a more scientific way and enumerated the diets to be given to patients in various diseases. But no one so far has written any considerable work on regimen in general, although this is a most important omission. Some of these authors were not unaware of the multiplicity of the different ways in which each disease may present itself; but they made mistakes when they tried to set down clearly the number of individual diseases. It is not easy to count accurately if a different name is given to every morbid condition differing but slightly from another; and unless a disease has the same name in all its forms it will appear to be a different disease.

4. I believe that attention should be paid to all the details of the science [of healing]. Measures requiring to be done well and exactly must be performed well and exactly; where speed is essential, with speed; where cleanliness is required, with cleanliness; and where pain is to be avoided, the patient should be treated so as to cause the minimum of pain. All such things should be done considerably better by the physician than by another.

5. I would single out for praise the physician who particularly excels in the treatment of acute diseases, for these cause the greatest number of deaths. By acute diseases are meant the conditions which earlier doctors have named pleurisy, pneumonia, brain-fever and *causus*, and conditions resembling them which usually show continued fever. For in the absence of an epidemic of a disease of the plague type, when the cases of illness are scattered, many more die of these conditions than all the others together.

6. Laymen, far from recognizing those who excel in the treatment of acute diseases, generally praise or blame any cure that is different. A good indication that the common people are at their most unintelligent in discussing these diseases, is that such cases give quacks their reputations as physicians. It is easy enough to learn the names of the things given to treat such patients, and if anyone talks of barley water, or of such and such a wine, or of hydromel, the layman thinks that all doctors, both good and bad, mean exactly the same thing. On the contrary, it is in such matters that their differences are clearly shown.

7. It seems to me worth while recording facts which in spite of their importance are not generally known to the medical profession, and to state what is harmful and what beneficial in the treatment of patients. For instance it is not generally known why some physicians all their lives give their patients gruel which is unstrained, thinking this is the right way to effect a cure; while others regard it as of the highest importance that the patient should not swallow a single grain of barley, as they think this causes great harm, but strain the barley-water through a linen cloth before giving it to their patients. Some

again give neither gruel nor barley-water; others give it only during the first seven days of the disease and yet others give it till the crisis be reached.

8. Physicians are quite unaccustomed to propound such questions, and perhaps they do not appreciate them when they are propounded. The science of medicine has fallen so low in popular estimation as not to seem the science of healing at all. As a result, if, in the acute diseases at least, practitioners differ so widely that the diet prescribed by one is regarded as bad by another, the science could almost be compared to divination. Seers think the same bird to be of good omen if it appears on the left and bad if it appears on the right, while other seers hold exactly the opposite view; and there are similar contradictions in divination by inspection of an animal's entrails.

9. I assert that this study of regimen is much to be recommended, and it is something closely allied to the most numerous and the most vital studies which compose the science of medicine. To the sick it is a powerful aid to recovery, to the healthy a means of preserving health, to athletes a means of reaching their best form and, in short, the means by which every man may realize his desire.

10. Barley-gruel seems to have been correctly selected as the most suitable cereal to give in these acute diseases and I have a high opinion of those who selected it. Its gluten is smooth, consistent and soothing; and is slippery and fairly soft; it is thirst-quenching and easily got rid of in case this be necessary. It contains nothing to produce constipation or serious rumbling, nor does it swell up in the stomach for during cooking it swells up to its maximum bulk.

11. Patients who take gruel in these diseases should not, as a general rule, fast on any day. They should take it without interruption unless the use of a purge or an enema renders a break necessary. Those who are accustomed to two meals a day should be given gruel twice daily; those who are accustomed to take only one meal a day should take gruel once only on the first day and thereafter it is permissible to increase this gradually and to give it twice a day if there seems to be any need for it. At the beginning of an illness it should be given

sparingly, nor should it be very thick; in fact, the patient should take only as much as he requires to allay an empty feeling.

12. If the disease is drier than one would like, the patient should be given a drink of either hydromel or wine – whichever is appropriate, and this will be discussed later – before the gruel, and increases in the quantity of gruel should be avoided. But if the mouth is moist and the pulmonary secretions are produced properly, the quantity of gruel should, as a general rule, be increased. For the sooner moist discharges appear and the more pronounced they are, the sooner will the crisis come, whereas delay in their appearance means that the crisis too will be delayed. Such is the general rule on these particular points.

13. There are many other important signs by which prognosis may be made; they will be passed over now to be treated later. The larger the stools, the more nourishment should be given until the crisis. This is specially so at the crisis and then extra large amounts should be given for the next few days. At least, this regimen should be adopted in those cases in which the crisis appears to take place on the fifth, seventh or ninth day; by so doing, precautions will have been taken for the following even day as well as the odd day of the crisis. Later, gruel should be given at first, giving place in time to solid food.

14. This treatment is generally successful if thick gruel is taken from the beginning. In cases of pleurisy, the pain stops spontaneously as soon as the patient begins to bring up any considerable amount of sputum and to be purged. Evacuation of discharges is much more complete and empyema less likely to occur on this regimen than if a different diet were taken. The crisis is more simple, more easily reached and less liable to be followed by a relapse.

15. The best barley should be used for gruel and it should be cooked as well as possible, especially if you intend to use only the barley-water. Apart from its other excellencies, the slipperiness of gruel makes the barley itself quite safe to swallow, for it does not adhere or lodge anywhere in its passage through the thorax. It is most slippery, thirst-quenching,

easily digested and weakest if it is really well cooked; all of which qualities are desirable.

16. A course of dieting on such gruel may be very harmful unless measures are taken to make it sufficient. Thus, if a patient has food retained in the stomach it will, unless he be made to evacuate some of it before being given gruel, only exacerbate any pain he already has, or give him one if he has none, and it will make respiration more rapid. This is bad because it dries up the lung and causes distress in the hypochondrium, the abdomen and diaphragm. Moreover, no gruel should be given to patients in whom the pain in the side be persistent in spite of warm fomentations, while the sputum is viscid and unripe and retained, unless the pain be relieved by relaxing the bowels or by cutting a vein, whichever may be indicated; if gruel is given to patients in such a condition, they will very quickly die.

17. For these and other similar reasons those who take thick gruel die within a week; in some cases after partly going out of their minds, in others choked by orthopnoea and stertorous breathing. It used to be thought that such patients had been the subject of a stroke, particularly because when they died the side was livid like a bruise. The reason for this appearance is that they die before the pain is allayed, for difficulty in breathing quickly sets in. Because the sputum becomes viscid and unripened, expiration is impeded causing wheezing in the bronchial tubes and thus, as has already been said, increased frequency of the respiration leads rapidly to asthma. When a patient reaches this condition, his case is generally desperate. The retained sputum actually prevents the intake of breath and forces it quickly to be expelled. Thus one thing is added to another. The retained sputum increases the rate of respiration and this in itself makes the sputum viscid so that it cannot run away. This may happen not only as the result of the untimely taking of gruel, but much more so if the patient eats or drinks anything less suitable.

18. In most respects, the additional precautions to be observed are the same whether the gruel is taken thick or strained to make barley-water. If neither of them, but only

drink is taken, the treatment to be given is sometimes different. The general rules are as follows.

19. If the fever begins soon after the patient has taken a meal and the bowels have not been opened, whether it be accompanied by pain or not, the diet of gruel should be withheld until it is judged that the food has passed to the lower part of the intestines. Fluids should be given and oxymel is recommended if there is pain, hot in winter and cold in summer. If there is acute thirst, hydromel and water may also be taken. If there is pain or any of the dangerous signs appear, but only after the seventh day or if the patient be strong, then gruel should be given. Should there be no evacuation of food previously consumed after recent food has been taken, an enema should be given to patients who are strong and in the prime of life; patients who are too weak should be given a suppository unless the bowels are opened satisfactorily of their own accord.

20. There is one time both at the beginning and, indeed, throughout the illness when gruel should not be given and that is when the feet are cold. It is then specially important not to administer any fluids, as well as to withhold gruel. But when warmth descends to the feet, then it may be given. It must be remembered that this is a time of great importance in all diseases, and not least in acute diseases and those accompanied by fever. Barley-water especially and gruel too should not be given without accurate observation of the signs which have been mentioned.

21. A pain in the side, whether it appears at the beginning of the illness or at a later stage, should first be treated in the ordinary manner in an attempt to remove it by hot fomentations. The best type of fomentation is hot water in a skin or bladder, or in an urn of bronze or earthenware. For comfort, something soft should first be put against the side. It is also good to apply a large soft sponge which has been dipped in hot water and wrung out. The warm object should be protected on top as in this way it stays hot longer and this also prevents steam reaching the patient's nostrils, unless of course this is regarded as beneficial; there are occasions when it is

needed. Barley and vetch may also be used if mixed with a little vinegar, sharper than one would drink, to soften it. It is then heated and sewn up in bags which are applied. Bran may be used in the same way. For dry fomentations, salt or millet is best, baked in woollen bags; millet is light and comforting.

22. Such a softening process also removes aches that extend up to the clavicles. Bleeding is not so efficacious in relieving pain. If the distress is not relieved by hot fomentations, heating should not be long continued as this dries up the lungs and causes empyema. If the pain causes a heavy feeling spreading towards the clavicle or arm, or about the breast or above the diaphragm, the inner vein at the elbow should be cut and you should not be afraid of drawing a large quantity of blood until, instead of running clear and red, it becomes either much redder or turns livid; either of these may happen.

23. If the pain is below the diaphragm and does not seem to extend towards the clavicles, the belly should be softened with either black hellebore or purple spurge, adding to the black hellebore, parsnip, seseli, cummin, anise or some other fragrant herb, and to the purple spurge the juice of silphium. These are also similar in effect if mixed with each other. But black hellebore gives a better evacuation and one more likely to produce a crisis, while purple spurge is better for breaking up wind. Both stop pain as do many other purgatives, but they are the best of those I know. Purgatives administered in the gruel are also helpful, so long as they are not too unpleasant owing to bitterness or any other unpleasant taste, or owing to the size of the dose or colour or anything else that may make them distasteful.

24. When the patient takes the purge, he should immediately be given a quantity of gruel not noticeably less than that to which he is accustomed. It is however customary to give no gruel during the purging. When purging stops, then less gruel than usual should be given, the amount subsequently being increased so long as the pain remains alleviated and no other contrary indication appears.

25. If it is proposed to give only barley-water, my advice is the same. I believe it to be better to start giving gruel right

away than to empty the body and then start a diet of gruel on the third, fourth, fifth, sixth or seventh day, unless the disease has already reached a crisis within that period. In this case too the same preparations should be made as have been described.

26. Such then is my opinion about the administration of gruel. As regards the sort of drink a patient should take, the gist of what I am going to say is very much the same. I know that physicians do the exact opposite of what is correct; they all want to dry up their patients for two or three days or more at the beginning of their illness, and then start to administer gruel and fluids. Perhaps it seems reasonable to them that when a violent change takes place in the body, it should be countered by a change equally violent.

27. A change in regimen may have considerable beneficial effects, but the change must be made in the right way and with intention. It is also important that the diet administered after the change should be correct. Those on a diet of thick gruel would suffer most if the change were incorrectly made, but those who are receiving only drink and those who take only barley-water would also suffer, the latter least.

28. Lessons should be drawn from our experience of what diets are best for men in health. If sudden changes in various diets are found to make a great difference to healthy people, it is only to be expected that they will have a great effect in disease, and the greatest in the acute diseases. It is well known that a low diet of food and drink is on the whole a surer way to health than violent changes from one diet to another. Sudden changes will harm those who take two meals a day and those who take one, and make them ill. Likewise, people who have not made a habit of taking luncheon are at once made ill if they take it. Their body feels heavy and they are weak and sluggish. If on top of this they dine they get heartburn. In some cases, too, loose stools are passed because the belly has been subjected to an unaccustomed load having been used to drying up and not being twice filled and having twice to digest a meal.

29. It is helpful in such cases to compensate for the change. An unwanted luncheon should be followed by a sleep, just as

we go to bed for the night after dinner. In winter, care should be taken to avoid shivering; in summer to avoid being too warm. If sleep will not come, a slow prolonged stroll, with no stops, should be taken. Dinner should be dispensed with, or only a little taken of something which can do no harm. Still less should be drunk and nothing watery. Such a man would suffer still more if he were to eat to repletion three times a day; still more if he did so more often. There are, however, many people who take three good meals a day without any ill effects, if they are accustomed to it.

30. It is also true that those who have been accustomed to eating two meals a day become weak, ill, slack at all kinds of work and suffer from heartburn, if they miss luncheon. Their intestines feel unsupported, they pass warm pale green urine and their stools are dried up. Sometimes, in addition, the mouth becomes bitter, the eyes sunken, the temples throb and the extremities become cold. As a rule those who have had no luncheon are unable to eat any dinner, or if they do their bellies feel heavy and they sleep less soundly than if they had previously taken luncheon.

31. Seeing that such things can happen to healthy people as the result of half a day's change in diet, it seems best in sickness not to give more nor less than the patient is accustomed to.

32. If a man, who contrary to his usual habit took only one meal, were to fast all day and eat his usual dinner, it is probable that it would lie still more heavily on him, seeing that he felt ill through missing luncheon and found his dinner lie heavy on him. The longer the fast which was suddenly broken, the more he would suffer.

33. An unaccustomed fast may be compensated for as follows. Excessive cold, heat or fatigue should be avoided, for all these would occasion distress. Less dinner should be taken than is usual and this should consist of the moist rather than dry foods. Drink should not be watery, and should not be less than usual in proportion to what is eaten. The next day a light luncheon should be taken and then a gradual return made to a normal diet.

34. People suffer most from these changes in diet when they

are of bilious disposition. Those who are phlegmatic generally suffer least discomfort from fasting, and so suffer less from taking only one meal a day contrary to their normal habits.

35. This will be sufficient to demonstrate that the most violent changes affecting our natures and constitutions are the most productive of illness. One must not without good reason order severe fasting, nor give food when a disease is at its height and is accompanied by inflammation, nor must one make sudden changes in either direction.

36. Many other related points concerning the belly could be mentioned, to show how well it puts up with food and drink to which it is accustomed, even if this is not naturally good. On the contrary, it has difficulty in digesting food and drink to which it is unaccustomed even if they are not bad in themselves.

37a. It would scarcely appear remarkable if pain in the stomach were caused by taking an excess of meat, or by garlic or silphium, either the juice or the stalk, or anything of the kind which has an individually potent effect. But it is surprising to learn how much distress, trouble, wind and colic in the stomach is caused by eating barley-cakes when one is accustomed to bread; or how much heaviness and constipation bread can cause one accustomed to barley-cakes. It is surprising, too, what thirst and sudden fullness is caused merely by eating bread when it is still warm, owing to its drying nature and the slowness with which it passes. Similarly, differing effects are produced by bread which is over-milled or made of unsifted meal if eaten by one not accustomed to it, or barley-cakes that are too dry or too moist or too sticky. Again, new barley-meal may affect those not accustomed to it, or that which is old, those accustomed to new.

Similarly, a sudden change in habit in which wine is substituted for water as a beverage, or vice versa, or the substitution of watered for neat wine: one produces distension of the upper part of the belly and wind in the lower, the other causes throbbing of the veins, heaviness in the head and thirst. Again, a change from white to red wine, even though both are equally strong, can cause an upset. All these things can cause many disturbances in the body. It would therefore appear less remarkable that a sudden change from a sweet to a strong wine, or vice versa, should fail to preserve a balanced constitution.

37b. I must however make a small concession here to the opposite school of thought. Such conditions are corrected by reversing the regimen, because changes of diet in these cases are not accompanied by changes in the body. The body is not growing stronger so as to need more food, nor weaker so as to need less.

38. The severity and character of each disease must be considered in relation to the patient and his customary diet both solid and liquid. Any increase is specially to be avoided since it is often advantageous to prescribe a total fast in those cases where the patient appears likely to be able to survive until the disease reaches its height. The cases in which this should be done will be described.

39. Much that is akin to what has been said might be added, but the following is the most convincing evidence. Not only is it related to the subject that forms my main topic, but it is itself a most opportune lesson. Those who are stricken with an acute disease sometimes eat food the very same day as the disease begins, some eat on the next day. Some swallow whatever is to hand, and some even drink *cyceōn.** All these possibilities are more harmful than if some other diet had been followed. However, mistakes are much less serious at this stage than they would be if the patient fasted totally for two or three days and then started eating on the fourth or fifth day. It would be still worse if he should fast on these days and then subsequently start to take these things before the disease had passed its height, for it is quite obvious that such a course is generally fatal unless the disease be extremely mild. But mistakes at the beginning are not so serious and are much easier to remedy. This is, I think, a most important lesson: during the first days of an illness the patient must not be forbidden any kind of gruel if gruel or solid food is shortly afterwards to be prescribed.

40. There is in fact utter ignorance among those who take

*A mixture of wine, cheese and barley.

barley-gruel that it is harmful if they have fasted for two or three days previously, and those who take barley-water are unaware that it too may do them harm unless they start taking it in the right way. It is however known, and the point is carefully observed, that it is very bad for the patient to drink barley-gruel before the disease has reached its height if he has been accustomed to barley-water.

41. All these things constitute clear evidence that physicians mishandle their patients' diets. They prescribe fasting in those diseases in which patients are going to be given gruel and who should not be prepared by fasting. They prescribe a change from fasting to gruel in just those cases in which a change should not be made. For the most part they prescribe the change from fasting to gruel at exactly the stage when it is beneficial to reduce the diet even to a complete fast, that is, when the disease is approaching a paroxysm.

*

43.* I observe also that physicians are not acquainted with the way in which one can distinguish the various causes of weakness during the course of an illness: which is due to fasting, which to some other provocation, which to distress and to the violence of the disease; nor can they distinguish the various states and appearances engendered by the constitution and condition of each one of us. Yet life or death may hang upon the ability to distinguish and to recognize such things.

44. It is in fact a serious fault to give a patient who is weak from distress and the violence of the disease more drink or gruel or solid food under the impression that his weakness is due to fasting. It is also an outrage to fail to realize that a patient's weakness is due to fasting and to make him worse by prescribing abstinence. Not only is this latter mistake dangerous, though less so than the former, but it is more likely to involve the physician in ridicule. For another physician or even a layman has only to come along and, having recognized what has happened to the patient, give him something to eat

*Section 42, which is clearly an interpolation from another work, is omitted.

and drink in defiance of the other's orders, and the error is plain to all to see. Such are particularly the occasions which expose the practitioner to ridicule since the physician or layman who intervenes seems almost to have raised a man from the dead. The signs by which each condition can be distinguished will therefore be described.

*

45.* This however is very similar to conditions in the belly. For if the whole body is rested much more than is usual, there is no immediate increase in strength. In fact, should a long period of inactivity be followed by a sudden return to exercise, there will be an obvious deterioration. The same is true of each separate part of the body. The feet and limbs would suffer in the same way if they were unaccustomed to exercise, or were exercised suddenly after a period of rest. The same is true of the teeth and of the eyes, and in fact of every part of the body. A softer bed than usual or one harder than usual causes distress, and sleeping in the open hardens the body.

46. A single illustration of all this will suffice. Suppose a man has a wound on the lower part of the leg which is neither very serious nor quite trifling, and not the sort which will heal very rapidly or very slowly. If from the first day he has it attended to and takes to his bed and never raises his leg, inflammation is less likely and he will be cured much more quickly than if he should walk about during treatment. If however on the fifth or sixth days, or even later, he should get up and walk about, he will suffer more distress than if he had been walking about from the beginning of the cure. And if at this stage he suddenly exerted himself much, he would suffer much more than if he had followed the other course of treatment and exerted himself to the same extent. All these facts hang together and constitute a proof that any change much in excess of what is moderate is harmful.

47. To take an immoderate amount of food after a long fast does very much more harm to the belly than to fast from a hearty diet, and may be compared with the effect on the other

*Something appears to be missing here.

parts of the body of over-exertion after a long period of rest. Just as the body should be given a complete rest and idleness, and slackness follow a long period of strenuous effort, so should the belly be given a rest from full feeding as otherwise it will cause pain and distress throughout the body.

48. Most of what I have said relates to changes from one diet to another. This is generally useful information, but something in particular must be added about the change from fasting to the taking of gruel in acute illnesses and this change must be made according to the instructions I give. Moreover, gruel must not be given until the disease has ripened or some sign has appeared either in the intestines, indicating starvation or irritation, or in the hypochondrium. These signs I shall describe.

49. Severe insomnia makes food and drink harder to digest. On the other hand, a change in the other direction relaxes the body and brings languor and headache.

50. The various effects in acute illnesses of the different sorts of wine, sweet or strong, white or red, and of hydromel, water or oxymel, can be judged from the following indications.

Sweet wine is less likely to produce headache than is heavy wine, it has less effect upon the mind and, as regards the internal organs, it is more easily passed than the other but causes enlargement of the spleen and liver. It is most unsuitable for those with bitter bile for it makes them thirsty. It may cause wind in the upper part of the intestine, but it does not trouble the lower part in this way. Wind caused by sweet wine does not easily escape but lingers about the hypochondrium. It is also, generally speaking, less easy to pass in the urine than is strong white wine. Sweet wine produces more sputum than the other kind. If one finds that drinking sweet wine causes thirst, it does not produce so much sputum as the other kind of wine; if it does not cause thirst, the opposite is true.

51. The main points in favour of and against white strong wine have already been pointed out in the description of sweet wine. As it passes more easily to the bladder than the other kind and is diuretic and purgative, it is always very beneficial in acute diseases. For even though it is less suitable than the

sweet in other respects, yet the cleansing through the bladder which it causes is beneficial so long as it is administered correctly. These are good points to note about the beneficial and the harmful properties of wine; they were unknown to my predecessors.

52. Tawny wine and bitter red wine should be employed in these diseases in this way. If there is neither headache nor affection of the mind, if there is no retention of sputum or urine, and if the stools are rather too loose and full of shreds, it is desirable to change from white or such wines to these. It should also be understood that the more it is diluted, the less harm it will do to the upper organs and to the bladder, while the less it is diluted, the greater is the benefit to the intestines.

53. To drink a mixture of honey and water throughout an illness caused by an acute disease is generally less suitable for those with bitter bile and enlarged viscera than it is for those who have not these things. It is less productive of thirst than is sweet wine and it softens the lung allowing sputum to be brought up in moderation and soothes a cough. It has some detersive quality which makes the sputum less tenacious than it would otherwise be. Hydromel is also a fair diuretic provided that none of the viscera interfere with this action. It also causes the passage of bilious stools, sometimes good ones, but sometimes they are excessive and more frothy than they should be. This is more likely to occur in those who are bilious or have enlargement of the viscera.

54. Softening of the lungs and expectoration of sputum is produced by a greater dilution of honey; frothy stools as well as those which are excessive and warmer than they should be, are due to a less diluted mixture. Stools of this kind bring other considerable troubles. Thus, instead of burning feelings in the hypochondrium being allayed they are provoked and cause distress, tossing of the limbs and ulceration of the internal organs or of the anus. Measures to prevent these things happening will be described.

55. The administration of honey and water without any gruel, instead of any other drink, is more often successful than not in these diseases. Of the reasons why in some cases it

should be given and in some not, the chief points have been stated.

56. Honey and water is generally acknowledged to enfeeble those who drink it and, for this reason, it has acquired a reputation for hastening death. It got this name from people starving to death, as some actually use this mixture for such a purpose. In fact it does not hasten death in all cases but is much more strengthening than water alone so long as it does not upset the stomach. Compared with white wine or with weak or odourless wine, it is in some ways more strengthening and in others more weakening. There is a vast difference in the effect on a patient's strength of wine and honey when taken undiluted. If a man were to eat a certain quantity of honey and another were to drink twice as much neat wine, the man who had eaten honey would gain much more strength so long as his stomach were not upset, for the wine causes the passage of much larger stools. If a man were to take gruel and then drink honey and water on top of it, the mixture would be too filling and would cause wind, besides being bad for the organs in the hypochondrium. However, it is not so harmful when drunk before gruel and may even be of some benefit.

57. Boiled hydromel is much more attractive in appearance than is the raw preparation, as it is then sparkling, thin, colourless and transparent. But I cannot attribute to it any other virtue that the raw drink does not possess. It is not even any sweeter than when taken raw, so long as the honey is good. It is, however, weaker and less productive of stools, neither of which are virtues in the case of honey and water. It is best used boiled when the honey is bad, not properly cleared, dark and of ill odour. Cooking will remove the worst of these faults.

58. You will also find the drink known as oxymel useful in these diseases as it promotes the bringing up of sputum and good respiration. The following are some useful points about it. If very sharp, its effect on tenacious sputum will be extreme. If it results in the bringing up of whatever is causing hoarseness and making the throat slippery and, as it were, sweeps the windpipe clean, then it will soothe the lungs owing to its

softening properties. Should all these things happen it is very beneficial. But sometimes for all its sharpness, oxymel fails to win the struggle to bring up the sputum but increases its viscosity and does harm. This is specially liable to happen in those who are in other respects likely to die and are unable to cough and fetch up the matter within. The patient's strength should be estimated with this in mind and, if there is hope, give it. If you do administer oxymel, give it just lukewarm, a little at a time and not in large quantities.

59. On the other hand, oxymel that is only slightly sharp moistens the mouth and pharynx, brings up the sputum and quenches thirst. It is good for the hypochondrium and for the neighbouring viscera. It also neutralizes the harmful effect of honey by correcting its bilious quality. It also breaks up wind and stimulates the passing of urine. However, it causes flabbiness in the lower part of the bowel and the passage of shreds. There are occasions when it is bad for those suffering from acute illnesses especially in that it prevents wind from passing through but makes it come back. It may also enfeeble the patient and chill his limbs. This is the only harm worth mentioning that oxymel can cause, so far as I know.

60. It is advisable to take a little of this drink of oxymel at night and on an empty stomach before taking gruel, though there is nothing to prevent its being taken a good while after the gruel is taken. Those who are subsisting on a completely fluid diet with no gruel will not find it suitable for continued use. This is chiefly because of the scraping and roughening which it produces in the intestines, and if the patient passes no stools it is likely to cause these things while the patient is taking nothing. Then, too, the honey and water might lose some of its strength. Should however the disease as a whole seem to benefit by the copious use of this draught, add only a suspicion of vinegar to the honey. This will avoid the most likely ill-effects and benefit the parts which need it.

61. To sum up, the sharpness obtained from vinegar is more beneficial to those with bitter bile than those with black bile because it dissolves bitter substances, turns them into phlegm and fetches them up. Black bile is lightened, brought up and

diluted, for vinegar brings up black bile. Vinegar is generally more harmful to women than to men as it may cause pains in the womb.

62. There is no virtue which I can attribute to the drinking of water in acute disease. It neither soothes a cough in pneumonia, nor does it promote the expectoration of sputum so well as other drinks, if taken throughout the illness. However, if a little water is taken when changing over from oxymel to hydromel, it brings up the sputum on account of the change in the quality of the drinks by causing a sort of flood. Otherwise it does not even quench thirst but rather causes bitterness because it is of bilious nature, and is thus bad for those of bilious constitution and for the hypochondrium. The worst time to drink it is on an empty stomach for it is then most bilious in its effect and weakening. Water also causes enlargement of the spleen and liver when these are inflamed and it distends the stomach causing indigestion. It passes through slowly because it is both cold and crude and promotes neither the passage of stools nor of urine. This naturally constipating effect may prove harmful. If ever drunk when the feet are cold, it does very great harm to any organ that it attacks.

63. If there is any suspicion of a violent headache or derangement of the mind in these diseases, wine must be completely avoided. In such a case water should be given or, if wine is taken, it should be well watered down and tawny and quite devoid of smell. After such a draught, a small quantity of water should be taken. This prevents the strength of the wine going to the head and affecting the mind. Instructions as to when water alone should be drunk, when in large quantities and when more moderately, when warm and when cold, have already been given in part. The remainder will be mentioned in the appropriate places.

64. Similarly, instructions will be given in dealing with each disease regarding the other drinks that may be taken. The correct indications will be given for giving drinks made from barley, herbs, raisins or the second pressing of grapes, from wheat, thistle or myrtle, pomegranates and the rest. The same applies to compounded drugs.

65. Bathing is beneficial to most patients, to some if used continuously, to others if intermittently. Sometimes it must be used less than one otherwise would owing to the patient's lack of adequate facilities. There are not many houses where the necessary equipment and servants of the right kind are available. A bath can do no little harm if it is not taken in the right way. A sheltered spot free from smoke is needed, and plenty of water. Baths should be frequent but not excessively so unless there is some special reason. It is better not to be rubbed with soap, but if soap-mixture be used it should be warm and added to the water in far larger quantities than is usually the case and a further generous quantity should be added later and more soon afterwards. The patient should not have far to go to the bath-tub and it should be easy to get in and out of it. The bather should be quiet and orderly and should do nothing for himself; others should pour the water and rub him. A large quantity of tepid water should be prepared and it should quickly be poured over the the bather. Sponges are better than scrapers and the body should not be allowed to get too dry before it is anointed. The head however should be dried as well as possible by wiping it with a sponge. Do not allow the extremities, the head and the rest of the body, to become chilled. Do not bathe shortly after food or drink, and do not eat or drink shortly after a bath.

66. The decision whether to bathe or not should rest largely with the patient if he is particularly fond of his bath and accustomed to it. Such people are more eager for it and derive benefit from bathing and suffer harm from abstaining. Generally speaking, it is most suitable for the treatment of cases of pneumonia and of *causus*, for bathing soothes pain in the side and chest and in the broad part of the back. It also causes the sputum to ripen and aids its expectoration; it promotes good respiration and relieves fatigue. It also relaxes the joints and softens the skin; it promotes the secretion of urine, cures headache and makes the nose moist.

67. Such then are the beneficial effects of bathing, if all necessities are available. Should however one or more of the necessary accompaniments be lacking, there is a risk that the

bath will do more harm than good, for any one thing can cause considerable harm if it is not previously prepared by the attendants as it should be.

Bathing is least opportune for those suffering from diseases in which the bowels are more relaxed than they should be, nor should patients bathe who are too constipated and have not had any evacuation first.

Baths are also bad for those who are weak, those suffering from nausea or from vomiting, those who are bringing up bilious matter and those who have epistaxis unless this is trivial; you know what is opportune. If the indications are slight, a bath should be taken whether it benefits the whole body or only the head.

68. If then the preparations are satisfactory and the patient welcomes the idea, a bath should be taken every day. It will do no harm to those who are fond of bathing if they take one twice a day. Bathing is much more suitable for those who are on a diet of whole gruel and not just barley-water, though there are times when the latter too may bathe. Least of all should those who take only fluids bathe, though here again it may be allowed in some cases.

A decision must be based on what has been said about the type of patients who will derive benefit or not, according to their regimen. Those who need any of the beneficial effects of bathing should bathe in so far as it benefits them; but in those cases where there is no need of these effects, and any of the signs which render bathing harmful is present, baths should be avoided.

APHORISMS

The anthology of medical truths which has been famous enough to add a word to the English language.

Section I

1. Life is short, science is long; opportunity is elusive, experiment is dangerous, judgement is difficult. It is not enough for the physician to do what is necessary, but the patient and the attendants must do their part as well, and circumstances must be favourable.

2. In disturbances of the stomach and when there is spontaneous vomiting, it is beneficial to the patient if the noxious matter be voided. If it is not, then the reverse is the case. Similarly with fasting; if the desired effect be obtained there is benefit, but otherwise it is harmful. Accordingly, the place and season, the age of the patient and the nature of the disease must all be considered.

3. In the case of athletes too good a condition of health is treacherous if it be an extreme state; for it cannot quietly stay as it is, and therefore, since it cannot change for the better, can only change for the worse. For this reason it is well to lose no time in putting an end to such a good condition of health, so that the body can start again to reconstitute itself. Do not allow the body to attain extreme thinness for that too is treacherous, but bring it only to a condition which will naturally continue unchanged, whatever that may be. Likewise fasting, if taken to extremes, is treacherous; and so also is putting on weight, if excessive.

4. A light and frugal dietary is dangerous in chronic complaints and in those acute diseases where it is not indicated. Dieting which causes excessive loss of weight, as well as the feeding up of the emaciated, is beset with difficulties.

5. Sick people are in error when they take a light diet which only increases their distress. Then, whatever be wrong, they

only become more ill on a light diet than they would on a slightly more substantial one. For this reason, light and frugal diets, when persisted in, are dangerous even for the healthy, because the undernourished do not bear an illness so well as the well nourished. Therefore, on the whole, light and frugal diets are more dangerous than those which are a little more substantial.

6. Desperate cases need the most desperate remedies.

7. During the specially acute phase of a disease, pain is most severe and the lightest possible diet is advisable. At other times, when a more substantial diet is permitted, it should be increased slowly in proportion as the seriousness of the disease decreases.

8. When the disease is at its height, then the lightest diet must be employed.

9. It must also be considered whether the patient will be strong enough for the diet prescribed when the disease is at its height. Will the patient be exhausted first and not be strong enough for the diet, or will the disease be blunted and exhausted first?

10. A light diet must be employed from the first in those diseases which rapidly approach their height. But when a disease only gradually attains its maximum severity, the diet need be reduced only then and for a little time before. Previously a richer diet may be employed depending upon the strength of the patient.

11. During a paroxysm the diet must be reduced, for an increase then would be harmful. Thus, in those diseases in which paroxysms occur at intervals, the diet must be reduced at each recrudescence.

12. Paroxysms and periods of remission may be foretold by the nature of the disease. Thus, the season of the year and the periodicity of the paroxysms, whether they be quotidian or tertian or at longer intervals, serve as indications. The signs which appear also assist. For instance, if sputum appear in a case of pleurisy early in the disease, it signifies that the illness will be a short one; if late, that the illness will be prolonged. The appearance of the urine, stools and sweat will also give

some indication of the expected duration and seriousness of a malady.

13. Old people bear fasting most easily, then adults, much less youths and least of all children. The more active they are, the less do they bear it.

14. Things which are growing have the greatest natural warmth and, accordingly, need most nourishment. Failing this, the body becomes exhausted. Old men have little warmth and they need little food which produces warmth; too much only extinguishes the warmth they have. For this reason, fevers are not so acute in old people for then the body is cold.

15. In winter and spring, stomachs are warm and sleep longest. Accordingly, more food should be given in these seasons, for the body produces more warmth and thus needs more nourishment. Young men and athletes show the truth of this.

16. Fluid diets are beneficial to all who suffer from fevers, but this is specially true in the case of children and those who are accustomed to such kind of food.

17. In deciding whether food should be given once or twice a day, more often or less, in greater or in smaller quantities at a time, one must consider habit, age, place and season.

18. Starchy food is most difficult to digest in summer and autumn, easiest in winter and next easiest in spring.

19. In those maladies where paroxysms occur at intervals, give no food just before the paroxysm nor compel the patient to take anything, but reduce his usual diet.

20. When a disease has attained the crisis, or when a crisis has just passed, do not disturb the patient with innovation in treatment either by the administration of drugs or by giving stimulants. Let them be.

21. The progress of a disease should be so guided, where guidance is needed, so that it develops in the most favourable manner according to its natural tendency.

22. Use drugs only when the disease for which you employ them has come to a head and not when it is developing, unless it be ripe for such treatment, which is rarely the case.

23. Do not judge the stools by their quantity but by their quality and the manner of them, what is needful and comfortable for the patient. Where it is necessary to bring the patient to a fainting condition, even this should be done, if he be strong enough to stand it.

24. In acute diseases employ drugs very seldom and only at the beginning. Even then, never prescribe them until you have made a thorough examination of the patient.

25. If what ought to be voided is voided, it is beneficial and easily borne by the patient; if not, it is borne with difficulty.

Section II

1. A disease in which sleep causes trouble is fatal. Where sleep is beneficial, it is not fatal.

2. Sleep that stops delirium is good.

3. Both sleep and wakefulness are bad if they exceed their due proportion.

4. Neither a surfeit of food nor of fasting is good, nor anything else which exceeds the measure of nature.

5. Unprovoked fatigue means disease.

6. Those who are suffering from a bodily malady and do not feel much of the pain of it, are also suffering from mental disease.

7. When bodies become thin over a long period of time, feed them up again slowly. But when the wasting has come on in a short time, feed them up again quickly.

8. If, subsequent to an illness, a patient does not derive strength from the nourishment he takes, it means he requires more food. But if this happens when adequate nourishment is taken, it means a purge is necessary.

9. When it is desired to purge, the aim should be an easy evacuation.

10. The more nourishment you give to a person who has not been purged, the more harm you do.

11. It is better to be full of drink than full of food.

12. What is left behind in the body after the crisis frequently causes relapse.

13. A patient finds the night before a crisis trying, but the succeeding night is generally more comfortable.

14. In the case of haemorrhage from the stomach, a change in the character of the stool, if not clearly unfavourable, may indicate a change for the better.

15. Examine the stool in the case of patients suffering from diseases of the throat or from tumours of the body. If it be bilious, the disease is part of a sickness of the whole body. But if it resembles a normal stool, then the disease is localized and it is safe to feed the body.

16. Hard work is undesirable for the underfed.

17. Over-eating causes sickness, as the cure shows.

18. Those who eat their food quickly in large pieces, quickly void it.

19. It is unwise to prophesy either death or recovery in acute diseases.

20. Those who have relaxed bowels when they are young have constipated ones in later life; but if their bowels are constipated in youth they become relaxed as they grow old.

21. Hunger is alleviated by the drinking of neat wine.

22. Disease which results from over-eating is cured by fasting; disease following fasting, by a surfeit. So with other things; cures may be effected by opposites.

23. Acute diseases attain the crisis within fourteen days.

24. In the progress of a disease, it is the fourth day in each period of seven days which is indicative. Taking the eighth day of the disease as the beginning of the second period of seven days, it is the eleventh day which must be observed since that is the fourth day of the second period. Again the seventeenth day must be watched since this is the fourth day after the fourteenth or the seventh after the eleventh.

25. In summer, quartan fevers are usually of short duration; in autumn they last long, especially those contracted when winter is near.

26. It is better that a fever should succeed a convulsion than a convulsion follow a fever.

27. Too much hope must not be put in the regression of a disease when this happens without obvious cause, neither

should deterioration occurring contrary to expectation be feared overmuch. Such changes are of uncertain significance and usually last but a short time.

28. In feverish illnesses it is bad either if the body remains superficially the same and does not waste, or if it wastes more than might be expected. In the former case the malady will be long; in the latter there is evidence of the patient weakening.

29. Purge at the start of an illness if you think fit, but, when a disease is at its height, it is better to withhold such action.

30. Everything is at its weakest at the beginning and at the end, but strongest at its height.

31. It is a bad thing if a patient does not put on weight when he is being fed up after an illness.

32. As a general rule, if those who are poorly take their food well at first, but fail to put on weight, they finish by refusing food. On the other hand, if they firmly refuse food at first but take it later on, they make a good recovery.

33. In every illness, a healthy frame of mind and an eager application to victuals is good. The reverse is bad.

34. There is less danger from a disease which is proper to the nature, condition and age of a patient, or to the time of year, than if it be not proper to one of these.

35. In all maladies, those who are fat about the belly do best; it is bad to be very thin and wasted there. Purging may be dangerous in the latter case.

36. Those who are in unhealthy bodily condition are very liable to faint from the administration of purgative drugs, as do those who do not take the right food.

37. Those who are in good bodily condition are hard to purge.

38. With regard to food and drink, it is better to take something slightly less suitable but pleasing than something more suitable but less pleasing.

39. The old feel ill less often than the young, but when they contract chronic ailments these usually accompany them to the grave.

40. Hoarseness and running of the nose do not 'ripen' in the very old.

41. Those who are subject to frequent and severe fainting attacks without obvious cause die suddenly.

42. It is impossible to cure a severe attack of apoplexy and no easy matter to cure a mild one.

43. Those who have been strangled and who are unconscious but not yet dead will not recover if there is foam about the lips.

44. Sudden death is more common in those who are naturally fat than in the lean.

45. The chief factor in the cure of epilepsy in the young is change, especially that due to growing up, but seasonal change of climate, or change of place or mode of life are also important.

46. If a patient be subject to two pains arising in different parts of the body simultaneously, the stronger blunts the other.

47. Pain and fever are more marked while pus is forming than when it is formed.

48. Rest, as soon as there is pain, is a great restorative in all disturbances of the body.

49. Those who are used to bearing an accustomed pain, even if they be weak and old, bear it more easily than the young and strong who are unaccustomed.

50. What has become customary by long endurance is wont to give less annoyance than what is not customary, even if the former be more severe. But it may sometimes be necessary to produce a change to what is unaccustomed.

51. It is dangerous to disturb the body violently whether it be by starvation or by feeding, by making it hot or cold, or in any way whatsoever. All excesses are inimical to nature. It is safer to proceed a little at a time, especially when changing from one regimen to another.

52. If you apply all the regular treatment without getting the regular result, do not therefore change the treatment so long as your original diagnosis remains unchanged.

53. Those who have relaxed bowels, if they are young, tend to do better than those with constipated ones, but worse if they are getting old; for it is a general rule that the bowels become constipated with advancing years.

54. A heavy physique is noble and not unpleasing in the young; in old age it is awkward and less desirable than a smaller stature.

Section III

1. The changes of the seasons are especially liable to beget diseases, as are great changes from heat to cold, or cold to heat in any season. Other changes in the weather have similarly severe effects.

2. Some natures are naturally well-suited to summer and some to winter; others are ill-suited to one or the other.

3. Diseases vary in their relationships one with another; some are opposed, some are mutually agreeable. Similarly, certain ages are well- or ill-suited to certain seasons, places and regimens.

4. When cold and heat both occur on the same day at any time of the year, then you must expect those diseases commonly encountered in autumn.

5. South winds cause deafness, misty vision, headache, sluggishness and a relaxed condition of the body. When this wind is prevalent these symptoms occur in illnesses. The north wind brings coughs, sore throats, constipation, retention of urine accompanied by rigors, pains in the sides and breast. When this wind is prevalent such things will be encountered among the sick.

6. When the summer is spring-like in character, then expect much sweating in the course of fevers.

7. During periods of drought fevers are high. If the whole year be mainly dry, whatever the general climatic condition produced, expect similar illnesses.

8. When the weather is seasonable and the crops ripen at the regular times, diseases are regular in their appearance and easily reach their crisis. When the weather is irregular, diseases are irregular and their crises difficult.

9. It is in autumn that diseases tend to be most acute and most likely to prove fatal. The spring is the healthiest and least fatal time of year.

10. Autumn is worst for consumptives.

11. As regards the seasons: a dry winter with northerly winds followed by a wet spring and southerly winds produces acute fevers, ophthalmia and dysentery in the summer. This is specially true of women and those of a watery constitution.

12. On the other hand, a damp mild winter accompanied by southerly winds, followed by a dry spring in which the wind is from the north, tends to produce miscarriage on the slightest pretext in women approaching term in the spring. If parturition is accomplished the children are weak and sickly, so that either the children die at once or, should they survive, they are thin and fall ill frequently. This same character of the seasons gives rise to dysentery and dry ophthalmia as well, while the aged suffer from catarrh which may speedily prove fatal.

13. A dry summer accompanied by northerly winds and a wet autumn with southerly winds produce during the following winter headaches, coughs, hoarseness, running at the nose and, in some cases, wasting.

14. Alternatively, a rainless autumn in which the winds come from the north is advantageous to women and those of a watery constitution. Others suffer from dry ophthalmia, from acute fevers, from running at the nose and, in some cases, from melancholy.

15. As regards the weather in general: drought is more healthy than rain and less likely to provoke fatal illness.

16. The diseases usually peculiar to rainy periods are chronic fevers, diarrhoea, gangrene, epilepsy, apoplexy and sore throats. Those peculiar to a time of drought are consumption, ophthalmia, arthritis, strangury and dysentery.

17. As for the daily changes in the weather: a north wind stimulates the body and makes it of good tone and agile, and makes for a good complexion and acuity of hearing; the bowels are constipated and the eyes sting. But a pain in the chest is made worse by such a wind. On the other hand, south winds relax the body, make the tissues moist, reduce acuity of hearing and produce headaches and vertigo. Movement both of the eyes and of the body generally is sluggish and the bowels relaxed.

18. As for the seasons, in spring and full summer children and young people do best; in summer and, up to a point, autumn, the old; while the winter suits best those between these two groups.

19. Every disease occurs at all seasons of the year but some of them more frequently occur and are of greater severity at certain times.

20. For example, madness, melancholy, epilepsy, haemorrhages, sore throats, catarrh, hoarseness, coughs, leprosy, vitiligo, ulcerative eruptions – these are very common – tumours and arthritis are all common in the spring.

21. In summer, while some of the foregoing occur, we must also expect continued fevers, *causus*, tertian fevers, vomiting, diarrhoea, ophthalmia, earache, ulcers in the mouth, gangrene of the genitalia and heat spots.

22. In autumn, while we still encounter many of the summer ailments, you must expect as well quartan fevers, irregular fevers, diseases of the spleen, dropsy, consumption, strangury, enteritis, dysentery, pains in the hips, sore throats, ileus, epilepsy, madness and melancholy.

23. During the winter season, pleurisy, pneumonia, lethargy, catarrh of the nose, hoarseness, cough, pain in the chest, pains in the side and loins, headache, vertigo and apoplexy all occur.

24. Then, if diseases be grouped according to different ages we find that new-born infants suffer from aphthae, vomiting, cough, insomnia, nightmares, inflammation of the umbilicus and discharging ears.

25. When teething takes place, we must add painful gums, fevers, convulsions and diarrhoea. These are specially to be expected during the eruption of the canines and in plump children or those with hard bellies.

26. As they grow older, tonsillitis, deflexions of the vertebrae of the neck, asthma, stone, infection with round worms and ascaris, pedunculated warts, priapism, scrofulous swellings in the cervical glands and other tumours are seen.

27. On approaching puberty, besides the foregoing diseases we must add long-continued fevers and epistaxis.

28. Usually children's diseases reach the crisis either in forty

days, in seven lunar months or in seven years. Others resolve on the approach of puberty. However, should a disease persist after puberty or, in the case of girls, the time when menstruation is established, it is likely to become chronic.

29. In youths, haemoptysis, consumption, acute fevers and epilepsy besides other ailments must be added, but especially those mentioned above.

30. Later, we encounter asthma, pleurisy, pneumonia, lethargy, inflammation of the brain, *causus*, chronic diarrhoea, cholera, dysentery, enteritis and haemorrhoids.

31. In the old, dyspnoea, catarrhal coughs, strangury, dysuria, arthritis, nephritis, dizziness, apoplexy, cachexia, pruritus of the whole body, insomnia, ascites and fluid in the eyes and nostrils, failing sight, blindness from glaucoma and deafness.

Section IV

1. Drugs may be administered to pregnant women, if required, from the fourth to the seventh month of gestation. After that period, the dose should be less. Care must also be exercised in giving drugs to infants and children.

2. Drugs should be used to evacuate from the body such substances which, should they flow of their own accord, would be beneficial. Those substances, evacuation of which would not be advantageous, should be stopped from coming.

3. If those substances are purged which ought to be purged, it is beneficial and the patient bears it well; if the reverse, it is borne ill.

4. In summer-time, use drugs acting rather on the upper part of the bowel; in winter the lower part.

5. The administration of drugs is attended with difficulty at the rising of the Dog Star and shortly before.

6. Thin subjects who are prone to vomiting should be given medicine for the upper bowel, but reduce the dose in winter.

7. The well-covered, who are not prone to vomiting, should be given drugs for the lower bowel, but in this case avoid the summer.

8. In purging consumptive patients, employ only small doses.

9. The bowel should be treated in melancholics by the same reasoning applying the opposite treatment.

10. In very acute conditions, administer the required drugs on the same day as they are shown to be required. It is bad for such conditions to last long.

11. Patients suffering from colicky pains about the navel and aching in the loins develop distension unless the malady is dispersed by drugs or by other means.

12. It is bad to administer drugs acting on the small bowel during the winter in patients prone to enteritis.

13. Patients in whom purgation of the upper bowel is attended with difficulty should have their bodies moistened beforehand by administering more food and giving more rest, before the prescription of hellebore.

14. When anyone takes a draught of hellebore, he should be made to move about rather than left to rest and sleep. Sea travel demonstrates the efficacy of movement in producing a disturbance of the intestines.

15. If you wish hellebore to act more efficiently, keep the patient moving. When you wish to stop its action, order rest and sleep.

16. Hellebore is a dangerous drug for those with healthy flesh since in these it induces convulsions.

17. A patient without fever and with no appetite who suffers from heartburn, vertigo and bitterness in the mouth requires medicine for the upper part of the body.

18. Pain above the diaphragm indicates the need for drugs acting on the upper part of the body; pain below, for those acting on the belly.

19. When a purge is given to a patient who is not thirsty, its action continues until he becomes thirsty.

20. If a patient without fever suffers from colic, heaviness of the legs and aching in the loins, he needs drugs for the lower organs.

21. Black excrement, like blood, appearing spontaneously has a serious significance whether it be accompanied by fever

or not. The darker it is the more serious the condition. But when dark stools are due to drugs, however dark the colour, it is of little significance.

22. The vomiting or passage of dark bile at the beginning of an illness is fatal.

23. Those who show great wasting, either from acute or chronic illness or from wounds, and then pass dark bile or something resembling black blood, die the next day.

24. Dysentery starting with the passage of black bile is fatal.

25. The vomiting of blood of any kind is bad; its passage as excrement is not a good sign, nor is the passage of black stools.

26. Cases of dysentery in which pieces resembling solid tissue appear in the stools are fatal.

27. If a fever be attended with considerable haemorrhage from any part of the body, the patient's bowels become relaxed during recovery.

28. Biliousness of the stool ceases upon the supervention of deafness, deafness upon the appearance of bilious excrement.

29. A rigor occurring on the sixth day of a fever is a sign of a dangerous crisis.

30. A paroxysm which appears at the same hour on one day as it departed on the previous day is a sign of a dangerous crisis.

31. Suppurative inflammations about the joints, especially about the jaws, may follow exhaustion from fevers.

32. During recovery from an illness, pain about a part indicates that suppuration will occur there.

33. But should pain in a part have existed before the onset of the disease, then that is the site where the malady establishes itself.

34. Sudden choking without swelling of the throat in a patient with fever leads to a fatal outcome.

35. If, in a patient suffering from a fever, the neck be suddenly twisted round and swallowing becomes almost impossible though there is no swelling, then he will die.

36. Paroxysms of sweating in the course of fevers occurring on the 3rd, 5th, 7th, 9th, 11th, 14th, 17th, 21st, 27th, 31st, and

34th days of the disease are of good omen. Such paroxysms mark the crisis of the disease. But should such paroxysms not occur, then expect pain, a long illness and relapse.

37. In severe fevers, cold paroxysms of sweating indicate death; in milder cases a long illness.

38. The appearance of sweat on a particular part of the body indicates disease in that part.

39. Should one part of the body be hotter or colder than the rest, disease is present in that part.

40. Changes from hot to cold and then to hot again, affecting the whole body, or changes of colour signify a long illness.

41. Severe sweating after sleep without obvious cause signifies that the body has too much nourishment. If it happens that the patient is not taking his food, then he needs purging.

42. Continued sweating, whether hot or cold, indicates disease. If the sweat be cold, it is a major illness; if hot, a minor.

43. Continued fevers are dangerous if they grow worse every other day; should they however be remittent, in whatever fashion, it means that there is no danger.

44. Prolonged fevers may give rise to swelling and pains in the joints.

45. Those suffering from swelling and pain in the joints as a result of fever are taking too much food.

46. When a rigor supervenes on an unremitting fever when the patient has already been weakened, the outcome is fatal.

47. Livid, bloody, foul-smelling or bilious sputum supervening in cases of continued fever is of bad significance. However, if such expectoration remove the diseased matter, all may be well. A similar rule applies to the urine and to the stools. But unless separation occurs properly through these parts the outlook is poor.

48. Cold skin associated with a high internal temperature and thirst in a patient with continued fever is fatal.

49. In a continued fever, if the lip, eyebrow, eye or nostril be distorted; if the patient, being already weak, does not see or does not hear – if any of these things happen, death is at hand.

50. If, in a continued fever, respiration becomes difficult and delirium occurs, expect a fatal outcome.

51. Unless an abscess associated with fever discharge about the time of the first crises, a long illness is to be expected.

52. There is nothing strange in those suffering from fevers, or from other illness, deliberately weeping. But if they weep spontaneously, in spite of themselves, it is of more significance.

53. When, in fevers, the gums suppurate, the fever is increased.

54. In fevers of the type of *causus* where there is a frequent dry cough irritating slightly, thirst is not produced.

55. Fevers lasting more than one day which follow on a bubo are all serious.

56. A paroxysm of sweating in the course of a fever which is not associated with a fall in temperature is of bad significance. It indicates excess of moisture in the body and the illness will be prolonged.

57. Fever, succeeding a convulsion or tetanus, ends the illness.

58. An attack of shivering supervening in a case of *causus* puts an end to it.

59. A pure tertian fever reaches its crisis after a maximum of seven paroxysms.

60. When in the course of a fever, deafness, epistaxis or disorder of the stomach supervenes, the illness is approaching its end.

61. If the length of a fever is not an odd number of days, relapse is likely to occur.

62. If jaundice appears in a case of fever in less than seven days, the outlook is bad unless watery discharges from the belly occur.

63. In fevers attended with daily rigors, the fever intermits daily; it is not remittent.

64. Jaundice occurring on the 7th, 9th, 11th or 14th day of a fever is favourable unless the right hypochondrium be hard. In other cases, the outlook is unfavourable.

65. A sensation of burning in the belly and heartburn are of bad significance in fevers.

66. In acute fevers convulsions or violent pains in the intestines are of bad significance.

67. In cases of fever, fear on waking from sleep, or convulsions, are serious.

68. Irregular breathing in cases of fever is bad since it indicates a fit.

69. When the urine of a man with fever is thick, full of clots and of small quantity, an increase in quantity and clarity is advantageous. Such a change is especially likely to occur if, from the beginning or very shortly afterwards, the urine has a sediment.

70. Those whose urine during a fever is turbid like that of a beast of burden either suffer from headache or will do so.

71. When the crisis of an illness is reached on the seventh day, the urine shows a red cloud on the fourth day and is otherwise normal.

72. Colourless urine is bad; it is specially common in those with disease of the brain.

73. If pain in the loins and fever supervene when the hypochondrium is distended and full of rumblings, the bowels become relaxed unless wind breaks or the patient passes a large quantity of urine.

74. When suppuration is suspected in a joint, the suppuration is avoided if the urine which flows is thick and white, like that which is seen sometimes in wearisome quartan fevers. If there is epistaxis as well, it very quickly stops.

75. Blood or pus in the urine indicates ulceration of the kidneys or of the bladder.

76. Small fleshy objects, the shape of hairs, in the urine which is thick, mean there is a discharge from the kidneys.

77. Thick urine containing bran-like particles indicates inflammation of the bladder.

78. The sudden appearance of blood in the urine indicates that a small renal vessel has burst.

79. A sandy urinary sediment shows that a stone is forming in the bladder.

80. When blood clots in the urine are accompanied by

strangury, abdominal and perineal pain, it is the parts about the bladder which are affected.

81. The presence of blood, pus and flakes in a foul-smelling urine indicates an ulcer of the bladder.

82. If a tumour form in the urethra, suppuration and discharge of the abscess produce resolution.

83. Much urine passed at night means a small stool.

Section V

1. Convulsions following the administration of hellebore are fatal.

2. Convulsions succeeding upon a wound are fatal.

3. When a convulsion or a fit of hiccoughs follows severe haemorrhage, the outlook is bad.

4. A convulsion or a fit of hiccoughs following excessive purging is bad.

5. If a drunk man suddenly becomes speechless in a fit, he will die after convulsions unless a fever ensue or unless, upon recovering from his hangover, he regains his voice.

6. Unless those who contract tetanus die within four days they recover.

7. Those who suffer from epilepsy in childhood recover from it, but when it first appears after the age of twenty-five it usually continues till death.

8. If sufferers from pleurisy do not cough up material within fourteen days, the inflammation produces empyema.

9. Consumption occurs most frequently between the ages of eighteen and thirty-five.

10. Those in whom, from a sore throat, the disease passes to the lungs, either die within seven days or, surviving this period, suffer from empyema.

11. If the sputum of those suffering from consumption have an unpleasant smell when poured on hot coals and if the hairs fall from the head, a fatal outcome results.

12. In consumption, loss of the hair of the head accompanied by diarrhoea is fatal.

13. Frothy blood comes from the lungs.

14. Diarrhoea supervening on consumption is a fatal sign.

15. If those patients in whom pleurisy has resulted in empyema evacuate the abscess by expectoration within forty days following its bursting, they recover. If this is not so, they become consumptive.

16. Frequent over-heating of the body causes the following troubles: relaxation of the flesh, nervous weakness, benumbing of the mind, haemorrhage, fainting attacks and in some cases death.

17. Cold causes fits, tetanus, gangrene and feverish shivering fits.

18. Cold is bad for the bones, teeth, nerves, brain and the spinal cord; heat is good for these structures.

19. Parts that have been chilled should be thoroughly warmed unless there is bleeding or the likelihood of this.

20. In patients already suffering from ulcers, cold is of severe effect. It hardens the flesh round about, causes pain without suppuration, gangrene, feverish rigors, spasms and tetanus.

21. In young, well nourished people, tetanus is sometimes seen which does not follow upon an ulcer. In these a cold douche in summer produces a reaction of warmth and this warmth effects a cure.

22. Warmth which produces suppuration, as it may do with some but not all ulcers, is an important sign of recovery. It softens the skin, dries it up and relieves pain. Further, rigors, spasms and tetanus are resolved. To the head, it relieves headaches. Warmth is also of value in the treatment of broken bones, especially when the bone is exposed. It is useful in the treatment of ulcers on the head. Those parts of the body where ulcers or gangrene have been caused by exposure to cold benefit much from warmth and a crisis is attained. Similar advantage from warmth is observed in cases of ulceration of the anus, the private parts, the womb and the bladder. Cold, in these diseases, is inimical and killing.

23. Cold should be applied in the following cases: when there is haemorrhage or the danger of one. In such cases apply the cold not to the actual spot from which bleeding occurs or

is expected, but round about. Cold should also be applied to boils or pustules when these tend to be red or suffused with fresh blood. Cold turns pustules dark when they are long-standing. Cold applications are also indicated in cases of erysipelas where there is no ulceration; if ulcers have formed it is harmful.

24. Cold substances such as snow and ice are harmful to the chest; they cause cough, haemorrhage and fluxes.

25. Swelling and pain in the joints unassociated with ulceration, gout and spasms, are mostly relieved and reduced by cold douches and the pain thus dispelled. A moderate numbness relieves pain.

26. Water that is capable of quick heating and quick cooling is very light.

27. When a patient feels thirsty at night and has a great desire to drink and if afterwards he sleep, it is a good sign.

28. Aromatic vapour baths are useful in the treatment of female disorders and would often be useful for other conditions too if they did not cause headaches.

30.* Acute diseases are fatal to pregnant women.

31. Miscarriage follows blood-letting in pregnant women, especially if the foetus be large.

32. If a woman vomits blood, this ceases with the onset of menstruation.

33. It is a good sign if epistaxis occurs in a woman whose menstruation has stopped.

34. Frequent diarrhoea in a pregnant woman renders her liable to a miscarriage.

35. When a woman who is afflicted with hysteria, or who is in difficult labour, sneezes, it should be regarded as a good sign.

36. If the menses are not the proper colour and do not occur regularly, it is a sign that a purge is required.

37. If the breasts of a pregnant woman regress suddenly, it means she will have a miscarriage.

38. If, in a woman who is carrying twins, one breast becomes thin, a miscarriage will occur of one of the children.

*29 (=IV, 1) is here omitted.

If the right breast is affected, the male child will be lost; if the left, the female.

39. If a woman who is neither pregnant nor has given birth produce milk, her menstruation has stopped.

40. It is a sign of madness when blood congeals about a woman's nipples.

41. To know whether a woman be pregnant, administer a draught of hydromel on retiring when she has had no supper. If she suffers from colic in the stomach she is pregnant; if not, she is not pregnant.

42. A pregnant woman is of good complexion if the child be male; of ill complexion if the child be female.

43. If a pregnant woman have erysipelas of the womb she will die.

44. Pregnant women who are abnormally delicate have a miscarriage before the foetus becomes sizeable.

45. When women of medium build have miscarriages at two or three months without obvious cause, it is because the placenta is full of mucus. It is thus unable to hold the weight of the foetus which therefore is ejected.

46. When abnormally fat women do not conceive, it is because the omentum is pressing on the mouth of the uterus. Until they become thin, they do not become pregnant.

47. If suppuration in the womb spreads to the region round the hip joint, lint pledgets should be used to stop it.

48. A male foetus inclines to the right, a female to the left.

49. When a drug which causes sneezing is used to expel the after-birth, stop up the mouth and nostrils.

50. To restrain a woman's menstruation, apply the largest possible cupping-glass to the nipples.

51. During pregnancy the mouth of the womb is closed.

52. If much milk flows from the breasts of a pregnant woman it means that the foetus is weak; but if the breasts be dry, the foetus is healthy.

53. When abortion is threatened the breasts become lax. If the nipples should become hard again, there will be pain in the nipples or in the hip joints or in the eyes or in the knees and abortion will not take place.

54. If the mouth of the womb is hard, it inevitably closes.

55. Pregnant women who catch fevers, or who become very emaciated without obvious cause, either have difficult and dangerous labours or run the risk of miscarriage.

56. A spasm or an attack of fainting following on menstruation is bad.

57. When menstruation is excessive or is suppressed, this indicates disease of the womb.

58. Inflammation of the rectum and of the womb produce strangury, as also do suppurative conditions of the kidneys. Inflammation of the liver, however, causes hiccough.

59. If a woman has not conceived and you wish to determine whether conception be possible, wrap her up in a cloak underneath which incense should be burned. If the odour seems to pass through the body to the nose and mouth, then she is not sterile.

60. Menstrual bleeding which occurs during pregnancy indicates an unhealthy foetus.

61. When in a woman who is suffering from neither rigors nor fever menstruation is suppressed and she suffers from nausea, she is pregnant.

62. Women in whom the cervical os is cold and thick tend not to conceive easily. Similarly, a very moist os drowns and destroys the semen while an unusually dry and hot condition destroys the seed from lack of nourishment. Women who are free from these extremes are those who conceive best.

63. It is much the same with males. Either the general laxness of the body is such that the inner pressure of the wind is insufficient to eject the semen, or, on account of the thickness of the tissues, adequate moisture does not pass through. Also, cold may prevent it from heating sufficiently to collect where it should; or the same may happen through excessive heat.

64. Milk is not recommended for those who suffer from headaches. It is bad, too, for patients with fever, those whose bellies are distended and full of rumbling and those who are thirsty. It is bad also for patients with acute fevers in whom the stools are bilious, and for those who have lost much blood

in the stool. It is good for patients liable to consumption if they have not too high a fever. It should also be given in cases with prolonged low fever, where the patient is abnormally wasted, provided none of the above-mentioned contra-indications is present.

65. Those who suffer from wounds with swelling in addition are not particularly liable to spasms or madness. However, if the swellings suddenly disappear, spasms and tetanus occur where the lesions are situated posteriorly. If, however, the lesions are on the front of the body, madness, acute pain in the sides or suppuration follows. Where the swellings are particularly red, dysentery is seen.

66. If swelling does not occur as a result of serious deep wounds, the outlook is very bad.

67. Loose soft elastic swellings are not serious; hard indurated swellings are serious.

68. When there is pain at the back of the head, some help may be given by dividing the vessel which runs vertically in the forehead.

69. Rigors in women usually begin in the loins and pass through the back to the head. In men they tend to begin in the back rather than in the front of the body, for instance in the thighs or forearms. But men have a porous skin, as is shown by the hairs.

70. Those who catch quartan fevers are not at all liable to spasms. Should they previously have a spasm and then develop a quartan fever, then the spasms stop.

71. Those whose skin is dry and taut die without sweating. Those whose skin is relaxed and porous die sweating.

72. Jaundiced patients do not suffer especially from flatulence.

Section VI

1. In cases of chronic enteritis, the occurrence of heartburn, should it not have occurred before, is a good sign.

2. Those whose noses tend to run and whose semen is watery tend to be rather sickly. Those in whom the reverse is true are healthier.

3. Loss of appetite is bad for long-standing cases of dysentery and particularly when the disease is accompanied by fever.

4. Ulcers with a peeling edge are malignant.

5. Care must be taken to determine whether there are any striking points about the site of any pains complained of; whether they are in the side, in the breast or anywhere else.

6. Diseases of the kidneys and of the bladder are difficult to cure in the aged.

7. Of those pains and swellings which occur in the belly, those which are on the surface are less serious than those which are not.

8. Ulcers on the body are difficult to heal in dropsical patients.

9. Widespread exanthems are not accompanied by much itching.

10. Severe headaches are cured should there be a flow of pus, blood or fluid from the nostrils, mouth or ears.

11. It is good when haemorrhoids supervene on cases of melancholy or where there is renal disease.

12. Where long-standing haemorrhoids have been cured there is danger of dropsy or of wasting supervening unless one be left untreated.

13. Sneezing supervening on an attack of hiccoughs relieves that condition.

14. When, in a case of dropsy, water flows from the blood vessels into the abdominal cavity, the condition is relieved.

15. Unprovoked vomiting puts an end to long-continued cases of diarrhoea.

16. Diarrhoea supervening in cases of pleurisy or pneumonia is a bad sign.

17. It is good when sufferers from ophthalmia have diarrhoea.

18. Deep wounds of the bladder, brain, heart, diaphragm, of any of the delicate entrails, the stomach or liver are fatal.

19. Division of bone, cartilage, nerve, the delicate part of the jaw, or of the foreskin is not followed by growing and joining together again.

20. Haemorrhage into the abdominal cavity is necessarily followed by suppuration.

21. Varicose veins or haemorrhoids appearing in a case of madness put an end to it.

22. Abscesses extending from the shoulders to the elbows are cured by bleeding.

23. Patients with fear or depression of long standing are subject to melancholia.

24. Division of the delicate entrails is not followed by repair.

25. When erysipelas, beginning on the surface, extends deeply into the body it is bad. However, deep inflammation coming to the surface is good.

26. Trembling in cases of *causus* is cured by delirium.

27. Should surgery or cauterization in patients with ulcers or with dropsy result in the loss of a great quantity of pus or watery fluid, death invariably follows.

28. Eunuchs are not subject to gout, nor do they become bald.

29. Gout does not occur in women except after the menopause.

30. A youth does not suffer from gout until after sexual intercourse.

31. Pains in the eyes are cured by drinking neat wine, by bathing, by vapour baths, by bleeding or by the administration of certain drugs.

32. People who lisp are especially liable to prolonged diarrhoea.

33. Those who suffer from heartburn are not particularly liable to pleurisy.

34. Those who are bald do not suffer from varicose veins, while should someone who is bald develop such veins, then his hair grows again.

35. It is bad when people with dropsy develop a cough, but good if they have the cough before the dropsy starts.

36. Dysuria is cured by bleeding and the incision should be in the inner vein.

37. In a case of sore throat, the development of swellings on the outer aspect of the trachea is a good sign.

38. It is better not to treat those who have internal cancers since, if treated, they die quickly; but if not treated they last a long time.

39. Spasms are cured either by over-eating or by fasting. The same is true of hiccoughs.

40. Pain around the hypochondrium, unattended by inflammation, is relieved by the onset of fever.

41. The thickness of the tissues themselves may be responsible for the absence of the signs of suppuration when the abscess is a deep one.

42. In cases of jaundice it is a bad sign when the liver becomes hard.

43. Should a splenetic patient catch dysentery and this become chronic, dropsy or enteritis supervenes and he dies.

44. Those who suffer from anuria as a result of strangury die within seven days unless, a fever supervening, a sufficient flow of urine is re-established.

45. Ulcers lasting a year or longer cause the underlying bone to be eaten away and the resulting scars are depressed.

46. Those who develop a hump-back from asthma or from cough before reaching puberty, die.

47. In cases where such treatment is advantageous, bleeding or purging is more efficacious in the spring.

48. To catch dysentery is helpful to splenetic patients.

49. The inflammation in all cases of gout subsides within forty days.

50. Laceration of the brain is invariably followed by fever and bilious vomiting.

51. Those in health who are suddenly taken with headache, loss of voice and who show stertorous breathing die within a week unless a fever supervene.

52. Observe the appearance of the eyes during sleep. Should any of the white be visible when the eyelids are closed, provided this be not due to diarrhoea or the taking of a drug, it is a sorry symptom and exceedingly fatal.

53. Raving delirium which is accompanied by laughter is safer; that accompanied by seriousness is more dangerous.

54. Respiration characterized by a sobbing sound in acute febrile illnesses is a bad sign.

55. Gout is usually most active in spring and autumn.

56. In melancholic diseases, a flow of humours to one part of the body is dangerous in that either apoplexy, a fit, madness or blindness will follow.

57. Apoplexy usually occurs between the ages of forty and sixty.

58. If the mesentery protrude, it invariably rots away.

59. If, in some condition, the hip-joint is dislocated and subsequently reduced again, fluid is formed.

60. If, following chronic pains in the hip, the joint becomes dislocated, the leg wastes away and the patient becomes lame. This may be prevented by the use of the cautery.

Section VII

1. It is a bad sign in acute illnesses when the extremities become cold.

2. It is a bad sign when the flesh becomes livid in the neighbourhood of a diseased bone.

3. It is bad when vomiting is followed by hiccough and bloodshot eyes.

4. Shuddering succeeding on sweating is not good.

5. It is a good thing in cases of madness when dysentery, dropsy or an ecstatic state supervenes.

6. In prolonged illnesses, loss of appetite and unadulterated excreta are bad.

7. A shivering fit and delirium following excessive drinking are bad.

8. The bursting of a tumour internally is accompanied by faintness, vomiting and swooning.

9. It is bad when delirium or spasms follow haemorrhage.

10. It is bad when vomiting, hiccough, fits or delirium be observed in a case of ileus.

11. It is bad when pneumonia supervenes upon pleurisy.

12. It is bad when inflammation of the brain supervenes upon pneumonia.

13. Fits or tetanus complicating severe burns are bad.

14. Shock or delirium following a blow on the head is bad.

15. It is bad when purulent sputum follows haemoptysis.

16. The production of purulent sputum is followed by consumption. When the sputum ceases, the patient dies.

17. Hiccough is a bad sign in cases of hepatitis.

18. A convulsion or delirium following insomnia is bad.

18a. Trembling following lethargy is bad.

19. Inflammation of the tissues following exposure of bone is bad.

20. Erysipelas followed by gangrene or by suppuration is bad.

21. When haemorrhage follows ulcers which throb violently, it is bad.

22. Long-standing pain in the belly followed by suppuration is bad.

23. Unmixed stools followed by dysentery are bad.

24. Delirium follows the fracture of a bone if the ends are not in apposition.

25. A convulsion following the administration of a drug is fatal.

26. It is bad when, following violent abdominal pains, the extremities become cold.

27. In the case of pregnant women, straining at stool may bring about miscarriage.

28. Bones, cartilages and nerves, when divided, will not reconstitute themselves.

29. An attack of diarrhoea puts an end to illnesses attended by the production of white phlegm.

30. The frothiness of the stools in certain cases of diarrhoea is due to substances flowing down from the head.

31. The presence of particles like coarse meal in the urine of patients with fever signifies a long illness.

32. A bilious-looking sediment in a urine which is clear above signifies an acute illness.

33. When the urine shows a deposit, there is some violent disturbance in the body.

34. Bubbles appearing on the surface of the urine indicate disease of the kidneys and a prolonged illness.

35. A considerable oily scum on the surface of the urine indicates an acute disease of the kidneys.

36. When in diseases of the kidneys the above signs occur together with acute pain in the region of the spinal musculature, an external abscess should be expected so long as the pain is felt superficially. But when the pain is located deeply, then the abscess will be situated deeply.

37. The vomiting of blood is a sign of recovery so long as it is unattended by fever; with fever it is bad. Cure may be effected in the latter case by cooling and the use of astringents.

38. Catarrh of the thoracic organs proceeds to suppuration within twenty days.

39. Patients who complain of strangury and pain in the perineum and in the pubic region, and whose urine contains blood and clots, are suffering from disease in the area round the bladder.

40. If the tongue be suddenly paralysed or if any part of the body be similarly affected, that is a sign of melancholia.

41. It is bad to purge old people so much that hiccough be produced.

42. Unless a fever be due to bile, the pouring of a lot of hot water on the head will end the fever.

43. A woman is never ambidexterous.

44. When empyemata are opened by the cautery or by the knife and the pus flows pure and white, the patient survives. But if it be mixed with blood, muddy and foul-smelling, he will die.

45. Those who are cauterized or cut for suppurating conditions of the liver survive if the pus runs pure and white, for then the abscess is encysted; if however it runs like lees of oil, they die.

46. Pains in the eyes should be treated by the administration of a draught of neat wine, the application of warm douches and the letting of blood.

47. There is no hope when a patient suffering from dropsy develops a cough.

48. Strangury and dysuria are relieved by the taking of neat wine and deep bleeding.

49. The appearance of redness and swelling on the chest is a good sign in cases of sore throat. It means that the disease has turned outwards.

50. Mortification of the brain is commonly followed by death within three days, but if these be survived, recovery will follow.

51. Sneezing occurs when the brain becomes thoroughly heated or when the sinuses become thoroughly moistened or chilled. As a result the air within is pushed out and in so doing makes a noise because its exit is through a narrow passage.

52. Severe pains in the liver disappear if fever supervenes.

53. Where a patient benefits from bleeding, it is best to bleed in spring-time.

54. When phlegm is enclosed between the diaphragm and the stomach causing pain, and is then unable to burst either into the thorax or the belly, it is evacuated by the blood vessels to the bladder and so the disease is resolved.

55. When the liver is full of fluid and this overflows into the peritoneal cavity, so that the belly becomes full of water, death follows.

56. Distress, yawning and shuddering are cured by a draught of wine mixed with an equal quantity of water.

57. The pain due to tumours in the urethra is relieved when they suppurate and burst.

58. Commotion of the brain, from any cause, is inevitably followed by loss of voice.

59. A fit of choking occurring during the course of a fever where there is no swelling of the throat, and which results in the patient being unable to swallow except with great difficulty, is fatal.

59a. If in a patient suffering from a fever the neck be suddenly twisted round and swallowing becomes almost impossible though there is no swelling, then he will die.

60. Those with too much moisture in their flesh should be treated by starvation since starvation dries up the body.

61. Changes from hot to cold and then to hot again, affecting the whole body, or changes of colour signify a long illness.

62. Continual sweating, either hot or cold, means that there

is an excess of fluid in the body. In the strong this should be removed by inducing vomiting; in the weak by purging.

63. Continued fevers are dangerous if they grow worse every other day; should they however be remittent, in whatever fashion, it means that there is no danger.

64. Prolonged fevers are attended either by swellings or pains in the joints.

65. Those who suffer from swellings or pains of the joints as the result of fever are taking too much food.

66. If the same diet be given to a patient with fever as would be suitable for a healthy man, although it would strengthen the healthy it would cause suffering to the sick.

67. The urine must be observed to see how far it resembles that passed in health. The less it resembles healthy urine, the more diseased it is; the more it resembles it, the healthier it is.

68. Where a sediment composed, as it were, of particles forms in the stools after standing, it is necessary to give a purge. If you give gruel to the patient before having cleansed the belly you do harm and the more you give, the greater the harm.

69. Where the stools are undigested, this is due to black bile. The more pronounced this tendency, the more pronounced the disease.

70. Livid, bloody, foul-smelling or bilious sputum in cases of continued fever is bad. But if healthy sputum come away it is good. Similar principles apply to the intestines and the bladder, for when such substances remain in the body and are not evacuated it is bad.

71. Whenever it is desired to rid the body of unwanted substances, an easy evacuation is desirable. If you wish to drain the thoracic organs, you must make the bowels costive; if you wish to drain the lower organs, you must make the bowels more relaxed.

72. Sleep and wakefulness, exceeding the average, mean disease.

73. Cold skin associated with a high internal temperature and thirst in a patient with continued fever is fatal.

74. If in the course of a continued fever a patient shows

distortion of the lip, nostril or eye or if the patient, being already weak, lose his sight or his hearing, he will die.

75. Dropsy supervenes on white phlegm.

76. Dysentery succeeds diarrhoea.

77. Lientery succeeds dysentery.

78. A suppurative inflammation of bone succeeds caries.

79–80. The coughing up of blood is followed by consumption and the production of purulent sputum. After consumption comes a discharge from the head and after this, diarrhoea. Following diarrhoea the sputum is no longer produced and when this stops death supervenes.

81. If the urine and the stools are abnormal, or if the discharges of the flesh or any other part of the body are unusual, then a slight deviation means a slight illness, a serious disturbance a serious illness and a very serious change, death.

82. Those who contract inflammation of the brain when over forty never recover. Less risk is run by those at that age and in that state of bodily development when the disease is more usually expected.

83. If during an illness there is weeping voluntarily, it is well. But if weeping occurs in spite of oneself, it is bad.

84. Epistaxis is a bad sign in quartan fevers.

85. Fits of sweating are dangerous when they take place on days other than those of the crisis. They may be very violent, drops of sweat swiftly collecting on the forehead, and the sweat flows away from the body in cold and profuse streams. Such an attack of sweating must necessarily be attended with a violent illness, excessive pain and long-lasting distress.

86. A violent discharge from the bowels is a bad thing when it occurs during continued illness.

87. What drugs will not cure, the knife will; what the knife will not cure, the cautery will; what the cautery will not cure must be considered incurable.

THE SACRED DISEASE

An attack on the popular superstitions about epilepsy, followed by an account of the natural history of the disease.

1. I do not believe that the 'Sacred Disease' is any more divine or sacred than any other disease but, on the contrary, has specific characteristics and a definite cause. Nevertheless, because it is completely different from other diseases, it has been regarded as a divine visitation by those who, being only human, view it with ignorance and astonishment. This theory of divine origin, though supported by the difficulty of understanding the malady, is weakened by the simplicity of the cure, consisting merely of ritual purification and incantation. If remarkable features in a malady were evidence of divine visitation, then there would be many 'sacred diseases', as I shall show. Quotidian, tertian and quartan fevers are among other diseases no less remarkable and portentous and yet no one regards them as having a divine origin. I do not believe that these diseases have any less claim to be caused by a god than the so-called 'sacred' disease but they are not the objects of popular wonder. Again, no less remarkably, I have seen men go mad and become delirious for no obvious reason and do many strange things. I have seen many cases of people groaning and shouting in their sleep, some who choke; others jump from their bed and run outside and remain out of their mind till they wake, when they are as healthy and sane as they were before, although perhaps rather pale and weak. These things are not isolated events but frequent occurrences. There are many other remarkable afflictions of various sorts, but it would take too long to describe them in detail.

2. It is my opinion that those who first called this disease 'sacred' were the sort of people we now call witch-doctors, faith-healers, quacks and charlatans. These are exactly the people who pretend to be very pious and to be particularly

wise. By invoking a divine element they were able to screen their own failure to give suitable treatment and so called this a 'sacred' malady to conceal their ignorance of its nature. By picking their phrases carefully, prescribing purifications and incantations along with abstinence from baths and from many foods unsuitable for the sick, they ensured that their therapeutic measures were safe for themselves. The following fish were forbidden as being the most harmful: mullet, black-tail, hammer and eel. Goat, venison, pork and dog were considered most likely among meats to upset the stomach. Of fowls: cock, turtle-dove and buzzard and those which are considered very rich were forbidden; white mint, garlic and onion were excluded from the diet because over-flavoured food is not good for a sick man. Further, their patients were forbidden to wear black because it is a sign of death, to use goat skin blankets or to wear goat skins, nor were they allowed to put one foot on the other or one hand on the other; and all these things were regarded as preventative measures against the disease. These prohibitions are added on account of the divine element in the malady, suggesting that these practitioners had special knowledge. They also employ other pretexts so that, if the patient be cured, their reputation for cleverness is enhanced while, if he dies, they can excuse themselves by explaining that the gods are to blame while they themselves did nothing wrong; that they did not prescribe the taking of any medicine whether liquid or solid, nor any baths which might have been responsible.

I suppose none of the inhabitants of the interior of Libya can possibly be healthy seeing that they sleep on goat skins and eat goat meat. In fact, they possess neither blanket, garment nor shoe that is not made of goat skin, because goats are the only animals they keep. If contact with or eating of this animal causes and exacerbates the disease while abstinence from it cures the disease, then diet is alone the factor which decides the onset of the disease and its cure. No god can be blamed and the purifications are useless and the idea of divine intervention comes to nought.

3. It seems, then, that those who attempt to cure disease by

this sort of treatment do not really consider the maladies thus treated of sacred or of divine origin. If the disease can be cured by purification and similar treatment then what is to prevent its being brought on by like devices? The man who can get rid of a disease by his magic could equally well bring it on; again there is nothing divine about this but a human element is involved. By such claims and trickery, these practitioners pretend a deeper knowledge than is given to others; with their prescriptions of 'sanctifications' and 'purifications', their patter about divine visitation and possession by devils, they seek to deceive. And yet I believe that all these professions of piety are really more like impiety and a denial of the existence of the gods, and all their religion and talk of divine visitation is an impious fraud which I shall proceed to expose.

4. If these people claim to know how to draw down the moon, cause an eclipse of the sun, make storms and fine weather, rain and drought, to make the sea too rough for sailing or the land infertile, and all the rest of their nonsense, then, whether they claim to be able to do it by magic or by some other method, they seem to be impious rogues. Either they do not believe in the existence of the gods or they believe that the gods are powerless or would not refrain from the most dastardly acts. Surely conduct such as this must render them hateful to the gods. If a man were to draw down the moon or cause an eclipse of the sun, or make storms or fine weather by magic and sacrifices, I should not call any of these things a divine visitation but a human one, because the divine power has been overcome and forced into subjection by the human will. But perhaps these claims are not true and it is men in search of a living who invent all these fancy tales about this particular disease and all the others too. They make a different god responsible for each of the different forms of the complaint.

If the sufferer acts like a goat, and if he roars, or has convulsions involving the right side, they say the Mother of the Gods is responsible. If he utters a higher-pitched and louder cry, they say he is like a horse and blame Poseidon. If the sufferer should be incontinent of faeces, as sometimes happens under the stress of an attack, Enodia is the name. If the stools are more frequent and thin like those of birds, it is Apollo Nomius; if he foam at the mouth and kick out with his feet, Ares is to blame. If he suffers at night from fear and panic, from attacks of insanity, or if he jumps out of bed and runs outside, they talk of attacks of Hecate and the assaults of the Heroes. In using purifications and spells they perform what I consider a most irreligious and impious act, for, in treating sufferers from this disease by purification with blood and like things, they behave as if the sufferers were ritually unclean, the victims of divine vengeance or of human magic or had done something sacrilegious. It would have been better if they had done the opposite and taken the sick into the temples, there, by sacrifice and prayer, to make supplication to the gods; instead they simply purify them and do none of these things. Charms are buried in the ground, thrown into the sea or carried off into the mountains where no one may touch them or tread on them. If a god really be responsible, surely these things should be taken into the temples as offerings.

Personally I believe that human bodies cannot be polluted by a god; the basest object by the most pure. But if the human body is polluted by some other agency or is harmed in some way, then the presence of a god would be more likely to purify and sanctify it than pollute it. It is the deity who purifies, sanctifies and cleanses us from the greatest and most unholy of our sins. We ourselves mark out the precincts of the temples of the gods so that no one should enter without purifying himself; as we go in, we sprinkle ourselves with holy water, not because we are thereby polluted, but to rid ourselves of any stain we may have contracted previously. This then is my opinion of the purifications.

5. I believe that this disease is not in the least more divine than any other but has the same nature as other diseases and a similiar cause. Moreover, it can be cured no less than other diseases so long as it has not become inveterate and too powerful for the drugs which are given.

Like other diseases it is hereditary. If a phlegmatic child is born of a phlegmatic parent, a bilious child of a bilious parent, a consumptive child of a consumptive parent and a splenetic child of a splenetic parent, why should the children of a father or mother who is afflicted with this disease not suffer similarly? The seed comes from all parts of the body; it is healthy when it comes from healthy parts, diseased when it comes from diseased parts. Another important proof that this disease is no more divine than any other lies in the fact that the phlegmatic are constitutionally liable to it while the bilious escape. If its origin were divine, all types would be affected alike without this particular distinction.

6. So far from this being the case, the brain is the seat of this disease, as it is of other very violent diseases. I shall explain clearly the manner in which it comes about and the reason for it.

The human brain, as in the case of all other animals, is double; a thin membrane runs down the middle and divides it. This is the reason why headache is not always located in the same site but may be on either side or, sometimes, affects the whole head. There are a large number of tenuous veins which extend to this structure from all parts of the body; there are also two large vessels, one coming from the liver and one from the spleen. That which comes from the liver is disposed as follows: one half runs down on the right side in relation with the kidney and the lumbar muscles, to reach the inside of the thigh and thence continues to the foot. It is called the 'hollow vein'. The other half courses upwards through the right side of the diaphragm and lies close to the right lung; branches split off to the heart and to the right arm while the remainder passes up behind the clavicle on the right side of the neck and there lies subcutaneously so as to be visible. It disappears close to the ear and then divides; the larger part finishes in the brain while smaller branches go separately to the right ear, the right eye and to the nostril. Such is the distribution of the blood-vessels from the liver. There is also a vein which extends both upwards and downwards from the spleen on the left side of the body; it is similar to that coming from the liver but is thinner and weaker.

7. It is through these blood-vessels that we respire, for they

allow the body to breathe by absorbing air, and it is distributed throughout the body by means of the minor vessels. The air is cooled in the blood-vessels and then released. Air cannot remain still but must move; if it remains still and is left behind in some part of the body, then that part becomes powerless. A proof of this is that if we compress some of the smaller blood-vessels when we are lying or sitting down, so that air cannot pass through the vessels, then numbness occurs at once. Such, then, is the nature of blood-vessels.

8. Now this disease attacks the phlegmatic but not the bilious. Its inception is even while the child is still within its mother's womb, for the brain is rid of undesirable matter and brought to full development, like the other parts, before birth. If this 'cleansing' takes place well and moderately so that neither too much nor too little comes away, the head is most healthy. But if there is too much lost from the whole brain so that a lot of wasting occurs, the head will be feeble and, when the child grows up, he will suffer from noises in the head and be unable to stand the sun or the cold. If the discharge is excessive from one part only, such as an eye or an ear, or one blood-vessel becomes shrivelled up, then whichever part be wasted in that way becomes damaged. On the other hand, if this 'cleansing' does not take place but the material is retained in the brain, a phlegmatic constitution is bound to result.

Sometimes phlegm, which should have been purged out during life in the womb, remains during early life and is only got rid of in the later years. This is what happens in the case of children who suffer from ulcers of the head, ears and flesh, and who salivate and discharge mucus; they get better as they grow older. Those who have been purged of the phlegm in this way are not troubled by this disease, but those who have neither been purged in this way by ulceration and discharges of mucus and saliva, nor have been purged in the womb, are liable to be attacked by it.

9. If these discharges should make their way to the heart, the chest is attacked and palpitation or asthma supervenes; some patients even become hump-backed. For when cold phlegm reaches the lungs and heart, the blood is chilled and the

blood-vessels, as a result of being violently cooled in the region of the lungs and heart, jump and the heart palpitates. Such circumstances force the onset of asthma and diseases characterized by orthopnoea because, until the phlegm which has flowed down has been warmed and dissipated by the blood-vessels, it is impossible to inspire as much air as is needed. When the phlegm has been removed, palpitation and asthma stop. The length of an attack depends upon the quantity of phlegm which has flowed in. The more frequent these discharges of phlegm, the more frequent the attacks. These effects, however, occur only if the discharge makes its way to the lungs and heart; if it reaches the stomach, diarrhoea results.

10. Should these routes for the passage of phlegm from the brain be blocked, the discharge enters the blood-vessels which I have described. This causes loss of voice, choking, foaming at the mouth, clenching of the teeth and convulsive movements of the hands; the eyes are fixed, the patient becomes unconscious and, in some cases, passes a stool. I will explain the reason for each of these signs. Loss of voice occurs when the phlegm suddenly descends in the blood-vessels and blocks them so that air can pass neither to the brain nor to the hollow blood-vessels nor to the body cavities, and thereby inhibits respiration. For when a man draws in breath through the mouth and nose, the air passes first to the brain and then the greater part goes to the stomach, but some flows into the lungs and blood-vessels. From these places it is dispensed throughout the rest of the body by means of the blood-vessels. The air which flows into the stomach cools it but makes no other contribution. But that which goes to the lungs and blood-vessels thence enters the body cavities and the brain and has a further purpose. It induces intelligence and is necessary for the movement of the limbs. Therefore, when the blood-vessels are shut off from this supply of air by the accumulation of phlegm and thus cannot afford it passage, the patient loses his voice and his wits. The hands become powerless and move convulsively for the blood can no longer maintain its customary flow. Divergence of the eyes takes place when the smaller blood-vessels supplying them are shut off and no longer provide

an air supply; the vessels then pulsate. The froth which appears at the lips comes from the lungs for, when air no longer enters them, they produce froth which is expectorated as in the dying. The violence of choking causes the passage of stools; choking is caused by the liver and the thoracic contents compressing the diaphragm and thus obstructing the entry into the stomach. This action results from the amount of air taken in by the mouth being less than normal. When air is shut off in the vessels of the limbs and cannot escape owing to the obstruction of the vessels with phlegm, it moves violently up and down through the blood and the convulsions and pain thus caused produce the kicking movements.

All these symptoms are produced when cold phlegm is discharged into the blood which is warm, so chilling the blood and obstructing its flow. If the cold material is copious and thick, the result is immediately fatal as though its coldness had overcome and congealed the blood. If the quantity is less, however, although at first it may have the upper hand and obstruct respiration, in the end it is dispersed throughout the blood which is plentiful and warm, and if it be overcome in this way, the blood-vessels again take in air and consciousness returns.

11. Infants who suffer from this disease usually die if the phlegm is copious and if the weather is southerly. Their little blood-vessels are too narrow to absorb a large quantity of inspissated phlegm and so the blood is at once chilled and frozen, thus causing death. If the amount of phlegm is small and enters both main vessels, or if it enters but one of them, the patient survives but bears the stigmata. Thus the mouth may be distorted, or an eye, a hand or the neck; according to the part of the body in which some blood-vessel became filled and obstructed with phlegm and thus rendered inadequate. As a result of this damage to the blood-vessel, the corresponding part of the body must necessarily be weakened.

Taking a long view such a happening is generally a good thing because a child is not liable to another attack after an attack which has produced some permanent damage. The reason for this is that the strain of the attack causes injury to and some narrowing of the remaining blood-vessels. As a result of this they will no longer admit the entry of phlegm to the same extent although they will admit air. It is, however, only to be expected that such deterioration in the condition of the blood-vessels will lead to some weakening of the limbs.

Those who have a very small discharge at a time when the weather is northerly recover without any permanent injury, but there is a danger in such cases that the disease will remain with the child as he grows older.

Such, then, is the way in which, more or less, this malady affects children.

12. Adults neither die from an attack of this disease, nor does it leave them with palsy. The blood-vessels in patients of this age are capacious and full of hot blood; as a result, the phlegm cannot gain the upper hand and chill and freeze the blood. Instead the phlegm is quickly overcome as it is diluted by the blood, and the vessels take in air again so that consciousness returns and the symptoms mentioned above are less pronounced owing to the strength of the patient.

Attacks of this disease in the aged are not fatal, nor do they cause paralysis. The reason is that the vessels are empty and the blood small in quantity and of thin and watery consistency. Nevertheless, a severe discharge of phlegm in winter may prove fatal if it takes place on both sides of the body by obstructing respiration and congealing the blood. If the discharge takes place on one side only, then, because the blood is too little, too cold and too thin, it cannot overcome the phlegm but instead is itself overcome and frozen. As a result, those parts of the body where the blood is destroyed become powerless.

13. The discharge of phlegm takes place more often on the right side of the body than on the left because the blood-vessels on that side are more numerous and of greater calibre than on the left.

The liquefaction and the subsequent discharge of phlegm occurs specially in children whose heads have been warmed thoroughly either by the sun or at a fire, and have then had the brain suddenly cooled, thus producing a separation of the phlegm. Although liquefying is produced by warmth and relaxation of the brain, it is the chilling and consolidating which makes the phlegm separate out, and thus causes the discharge. Such is the explanation in some cases; in others, after a period in which the wind has been in the north for some time and then shifts to the south, the brain which is consolidated and healthy becomes soft and relaxed so that there is an overflow of phlegm, thus causing a discharge. A discharge may also occur from obscure causes as when a patient has a fright, or is startled by someone shouting, or when sobs will not let him take in a breath quickly enough, as often happens with children. When such things happen the body immediately becomes cold and loss of voice is succeeded by apnoea. When breathing stops the brain congeals and the blood stops; thus the phlegm is secreted and discharged. Such are the causes of fits from which children at first may suffer.

In older people, the winter is the most dangerous time. When they get their heads and brains warm in front of a roaring fire and then go out and shiver in the cold, or when they come out of the cold into a warm room and a hot fire, the same thing happens for the reasons already given and they have a fit. There is also a grave risk of the same thing happening in the spring as a result of sun-stroke; it is least likely to happen in the summer when there are no sudden variations in temperature. Cases who have been free from the disease in childhood having their first attack after the age of twenty are very rare if not unheard of. At this time of life the vessels are filled with a great quantity of blood while the brain is stiff and solid. There is thus no discharge into the blood vessels or, if there is, it does not overcome the blood because this is so ample and warm.

14. When the disease has been present from childhood, a habit develops of attacks occurring at any change of wind and specially when it is southerly. This is hard to cure because the brain has become more moist than normal and is flooded with phlegm. This renders discharges more frequent. The phlegm can no longer be completely separated out; neither can the brain, which remains wet and soaked, be dried up.

This observation results specially from a study of animals, particularly of goats which are liable to this disease. Indeed, they are peculiarly susceptible to it. If you cut open the head you will find that the brain is wet, full of fluid and foul-smelling, convincing proof that disease and not the deity is harming the body. It is just the same with man, for when the malady becomes chronic, it becomes incurable. The brain is dissolved by phlegm and liquefies; the melted substance thus formed turns into water which surrounds the brain on the outside and washes round it like the sea round an island. Consequently, fits become more frequent and require less to cause them. The disease therefore becomes very chronic as the fluid surrounding the brain is dilute because its quantity is so great, and as a result it may be quickly overcome by the blood and warmed.

15. Patients who suffer from this disease have a premonitory indication of an attack. In such circumstances they avoid company, going home if they are near enough, or to the loneliest spot they can find if they are not, so that as few people as possible will see them fall, and they at once wrap their heads up in their coats. This is the normal reaction to embarrassment and not, as most people suppose, from fear of the demon. Small children, from inexperience and being unaccustomed to the disease, at first fall down wherever they happen to be. Later, after a number of attacks, they run to their mothers or to someone whom they know well when they feel one coming on. This is through fear and fright at what they feel, for they have not yet learnt to feel ashamed.

16. The reasons for attacks occurring when there is a change of wind are, I believe, the following. Attacks are most likely to occur when the wind is southerly; less when it is northerly, less still when it is in any other quarter; for the south and north winds are the strongest of the winds and the most opposed in direction and in influence. The north wind precipitates the moisture in the air so that the cloudy and damp elements are separated out leaving the atmosphere clear and bright. It treats similarly all the other vapours which arise from the sea or from other stretches of water, distilling out

from them the damp and dark elements. It does the same for human beings and it is therefore the healthiest wind. The south wind has just the opposite effect. It starts by vaporizing the precipitated moisture because it does not generally blow very hard at first. This calm period occurs because the wind cannot immediately absorb the moisture in the air which was previously dense and congealed, but loosens it in time. The south wind has the same effect on the earth, the sea, rivers, springs, wells and everything that grows or contains moisture. In fact, everything contains moisture in a greater or lesser degree and thus all these things feel the effect of the south wind and become dark instead of bright, warm instead of cold and moist instead of dry. Jars in the house or in the cellars which contain wine or any other liquid are influenced by the south wind and change their appearance. The south wind also makes the sun, moon and stars much dimmer than usual.

Seeing that such large and powerful bodies are overcome and that the human body is made to feel changes of wind and undergo changes at that time, it follows that southerly winds relax the brain and make it flabby, relaxing the blood-vessels at the same time. Northerly winds, on the other hand, solidify the healthy part of the brain while any morbid part is separated out and forms a fluid layer round the outside. Thus it is that discharges occur when the wind changes. It is seen, then, that this disease rises and flourishes according to changes we can see come and go. It is no more difficult to understand, nor is it any more divine than any other malady.

17. It ought to be generally known that the source of our pleasure, merriment, laughter and amusement, as of our grief, pain, anxiety and tears, is none other than the brain. It is specially the organ which enables us to think, see and hear, and to distinguish the ugly and the beautiful, the bad and the good, pleasant and unpleasant. Sometimes we judge according to convention; at other times according to the perceptions of expediency. It is the brain too which is the seat of madness and delirium, of the fears and frights which assail us, often by night, but sometimes even by day; it is there where lies the cause of insomnia and sleep-walking, of thoughts that will not

come, forgotten duties and eccentricities. All such things result from an unhealthy condition of the brain; it may be warmer than it should be, or it may be colder, or moister or drier, or in any other abnormal state. Moistness is the cause of madness for when the brain is abnormally moist it is necessarily agitated and this agitation prevents sight or hearing being steady. Because of this, varying visual and acoustic sensations are produced, while the tongue can only describe things as they appear and sound. So long as the brain is still, a man is in his right mind.

18. The brain may be attacked both by phlegm and by bile and the two types of disorder which result may be distinguished thus: those whose madness results from phlegm are quiet and neither shout nor make a disturbance; those whose madness results from bile shout, play tricks and will not keep still but are always up to some mischief. Such are the causes of continued madness, but fears and frights may be caused by changes in the brain. Such a change occurs when it is warmed and that is the effect bile has when, flowing from the rest of the body, it courses to the brain along the blood-vessels. Fright continues until the bile runs away again into the blood-vessels and into the body. Feelings of pain and nausea result from inopportune cooling and abnormal consolidation of the brain and this is the effect of phlegm. The same condition is responsible for loss of memory. Those of a bilious constitution are liable to shout and to cry out during the night when the brain is suddenly heated; those of phlegmatic constitution do not suffer in this way. Warming of the brain also takes place when a plethora of blood finds its way to the brain and boils. It courses along the blood-vessels I have described in great quantity when a man is having a nightmare and is in a state of terror. He reacts in sleep in the same way that he would if he were awake; his face burns, his eyes are bloodshot as they are when scared or when the mind is intent upon the commission of a crime. All this ceases as soon as the man wakes and the blood is dispersed again into the blood-vessels.

19. For these reasons I believe the brain to be the most potent organ in the body. So long as it is healthy, it is the

interpreter of what is derived from the air. Consciousness is caused by air. The eyes, ears, tongue, hands and feet perform actions which are planned by the brain, for there is a measure of conscious thought throughout the body proportionate to the amount of air which it receives. The brain is also the organ of comprehension, for when a man draws in a breath it reaches the brain first, and thence is dispersed into the rest of the body, having left behind in the brain its vigour and whatever pertains to consciousness and intelligence. If the air went first to the body and subsequently to the brain, the power of understanding would be left to the flesh and to the blood-vessels; it would only reach the brain hot and when it was no longer pure owing to admixture with fluid from the flesh and from the blood and this would blunt its keenness.

20. I therefore assert that the brain is the interpreter of comprehension. Accident and convention have falsely ascribed that function to the diaphragm* which does not and could not possess it. I know of no way in which the diaphragm can think and be conscious, except that a sudden access of pleasure or of pain might make it jump and throb because it is so thin and is under greater tension than any other part of the body. Moreover, it has no cavity into which it might receive anything good or bad that comes upon it, but the weakness of its construction makes it liable to disturbance by either of these forces. It is no quicker in perception than any other part of the body, and its name and associations are quite unwarranted, just as parts of the heart are called auricles though they make no contribution to hearing. Some say too that we think with our hearts and it is the heart which suffers pain and feels anxiety. There is no truth in this although it is convulsed as is the diaphragm and even more for the following reasons: blood-vessels from all parts of the body run to the heart and these connections ensure that it can feel if any pain or strain occurs in the body. Moreover, the body cannot help giving a shudder and a contraction when subjected to pain and the same effect is produced by an excess of joy, which heart and diaphragm

*Gk *phrenes* (diaphragm) is frequently used for 'mind' in the widest sense. The words for thinking, consciousness, etc., are closely connected.

feel most intensely. Neither of these organs takes any part in mental operations, which are completely undertaken by the brain. As then the brain is the first organ in the body to perceive the consciousness derived from the air, if the seasons cause any violent change in the air, the brain undergoes its greatest variations. This is my reason for asserting that the diseases which attack the brain are the most acute, most serious and most fatal, and the hardest problem in diagnosis for the unskilled practitioner.

21. This so-called 'sacred disease' is due to the same causes as all other diseases, to the things we see come and go, the cold and the sun too, the changing and inconstant winds. These things are divine so that there is no need to regard this disease as more divine than any other; all are alike divine and all human. Each has its own nature and character and there is nothing in any disease which is unintelligible or which is insusceptible to treatment. The majority of maladies may be cured by the same things as caused them. One thing nourishes one thing, another another and sometimes destroys it too. The physician must know of these things in order to be able to recognize the opportune moment to nourish and increase one thing while robbing another of its sustenance and so destroying it.

In this disease as in all others, it should be your aim not to make the disease worse, but to wear it down by applying the remedies most hostile to the disease and those things to which it is unaccustomed. A malady flourishes and grows in its accustomed circumstances but is blunted and declines when attacked by a hostile substance. A man with the knowledge of how to produce by means of a regimen dryness and moisture, cold and heat in the human body, could cure this disease too provided that he could distinguish the right moment for the application of the remedies. He would not need to resort to purifications and magic spells.

DREAMS

(REGIMEN IV)

This short treatise on the medical significance of dreams forms the conclusion to a long work on Regimen; this explains the numbering of the paragraphs, and the last sentence must be understood to refer to the whole work.

86. Accurate knowledge about the signs which occur in dreams will be found very valuable for all purposes. While the body is awake, the soul is not under its own control, but is split into various portions each being devoted to some special bodily function such as hearing, vision, touch, locomotion and all the various actions of the body. But when the body is at rest, the soul is stirred and roused and becomes its own master, and itself performs all the functions of the body. When the body is sleeping it receives no sensations, but the soul being awake at that time perceives everything; it sees what is visible, it hears what is audible, it walks, it touches, it feels pain and thinks. In short, during sleep the soul performs all the functions of both body and soul. A correct appreciation of these things implies considerable wisdom.

87. There are special interpreters, with their own science of these matters, for the god-given dreams which give to cities or to individuals foreknowledge of the future. Such people also interpret the signs derived from the soul which indicate bodily states; excess or lack of what is natural, or of some unusual change. In such matters they are sometimes right and sometimes wrong, but in neither case do they know why it happens, whether they are right or wrong, but nevertheless they give advice so you shall 'beware of taking harm'. Yet they never show you how you ought to beware, but merely tell you to pray to the gods. Prayer is a good thing, but one should take on part of the burden oneself and call on the gods only to help.

88. The facts about dreams are as follows: those that merely

consist of a transference to the night of a person's daytime actions and thoughts, which continue to happen in normal fashion just as they were done and thought during the day, are good for they indicate a healthy state. This is because the soul remains true to its daytime cogitations, and is overcome neither by excess nor by emptiness, nor by any other extraneous circumstance. But when dreams take on a character contrary to daytime activities and involve conflict or victory over them, then they constitute a sign of bodily disturbance. The seriousness of the conflict is an indication of the seriousness of the mischief. Now concerning this, I make no judgement whether or not you ought to avert the consequence by appropriate rites or not. But I do advise treatment of the body, for an excretion resulting from some bodily superfluity has disturbed the soul. If the opposing force be strong, it is a good thing to give an emetic and to administer a gradually increasing light diet for five days, to order frequent early-morning walks gradually becoming more brisk, and gymnastics for those accustomed to this form of exercise, proportionate in severity to the increase of diet. If the opposing force be weaker, dispense with the emetic, reduce the diet by a third and restore the cut by a gradual measure over five days. Strenuous walks and the use of vocal exercises will put an end to the disturbance.

89. It is a good sign to see the sun, moon, sky and stars clear and undimmed, each being placed normally in its right place, since it shows that the body is well and free from disturbing influences. But it is necessary to follow a régime which will ensure that such a condition is maintained. On the contrary, if any of these celestial bodies appear displaced or changed then such a sign indicates bodily disease, the severity of which depends upon the seriousness of the interference.

Now the orbit of the stars is the outermost, that of the sun is intermediate, while that of the moon is nearest to the hollow vault of the sky. Should one of the stars seem to be injured, or should it disappear or stop in its revolution as a result of mist or cloud, this is a weak sign. If such a change be produced by rain or hail, it is stronger and signifies that an excretion of moisture and phlegm has occurred into the corresponding outermost parts. In such cases, prescribe long runs well wrapped up, increasing the exercise so as to cause as much sweating as possible. The exercise should be followed by long walks and the patient should go without breakfast. Food should be cut by a third and the normal diet restored gradually over five days. If the disorder appears more severe, prescribe vapour baths in addition. It is advisable to cleanse through the skin because the harm is in the outermost parts. Therefore prescribe dry, pungent, bitter, undiluted foods and the most dehydrating exercises.

If the moon is involved, it is advisable to draw off the harmful matter internally; therefore to use an emetic following the administration of pungent, salty and soft foods. Also, prescribe brisk runs on a circular track, walks and vocal exercises. Forbid breakfast and reduce the food intake, restoring it as before. The cleansing should be done internally because the harm appeared in the hollows of the body.

If the sun encounters any of these changes, the trouble is more violent and less easy to expel. The drawing-off should be produced both ways; prescribe runs on the stadium track and on the circular track, walks and all other forms of exercise. Give an emetic, cut the food and restore the diet gradually over five days as before.

If the heavenly bodies are seen dimly in a clear sky, and shine weakly and seem to be stopped from revolving by dryness, then it is a sign that there is a danger of incurring sickness. Exercise should be stopped while a fluid diet, frequent baths and plenty of rest and sleep should be prescribed until there is a return to normal.

If the heavenly bodies are opposed by a fiery atmosphere, the excretion of bile is indicated. If the opposing powers get the upper hand, sickness is portended; but if they completely overcome the stars and these vanish, then there is danger that the sickness may terminate fatally. If the opposing influences, however, are put to flight and it seems as if they are pursued by the heavenly bodies, then there is danger of the patient going mad unless he be treated. In all these cases, it is best to start treatment by purging with hellebore. If this is not done, the diet should be fluid and no wine should be taken unless it be white, thin, soft and watery. Warm, pungent, dehydrating and salt things should be avoided. Prescribe as much natural exercise as possible and plenty of runs with the patients well wrapped-up. Avoid massage, wrestling and wrestling in dust. Soften them with plenty of sleep and, apart from natural exercise, let them rest. Let them take a walk after dinner. It is also good to take a vapour bath followed by an emetic. For thirty days the patient should not eat his fill, but when he is restored to a full diet he should take an emetic thrice monthly after partaking of a sweet, fluid and light meal.

When the heavenly bodies wander in different directions, some mental disturbance as a result of anxiety is indicated. In this case, ease is beneficial. The soul should be turned to entertainments, especially amusing ones, or failing these, any that may give special pleasure, for two or three days. This may effect a cure; if not, the mental anxiety may engender disease.

It is a sign of health if a star, which is clear and bright, appears to fall out of its orbit and to move eastwards. The separation of any clear substance and its natural excretion from the body is good. Thus excretion of substances into the bowels and the formation of abscessions in the skin are examples of things falling out of their orbit.

It is a sign of sickness if the star appears dark and dim and moves either westward, or down into the earth or sea, or upwards. Upward movement indicates fluxes in the head; movement into the sea, disease of the bowels; earthward movement, the growing of tumours in the flesh. In these cases it is wise to reduce the food intake by a third and, after an emetic, to increase it over five days. Then a normal diet should be taken for a further five days, after which another emetic should be taken followed by an increase in the same way.

It is a healthy sign if any of the heavenly bodies appears clear and moist, because the influx from the ether acting on the person is clear and the soul perceives this as it enters. If it be dark, and not clean and transparent, then sickness is indicated, not due to some internal excess or lack of something, but coming from the external environment. In this case it is advisable to take brisk runs on a circular track so as to restrict the wasting of the body. Also, the quickened respiration causes excretion of the intruding influence, and brisk walks should follow the runs. The diet should be soft and light, being increased to reach the normal in four days.

When a person appears to receive something pure from a pure deity, it is good for health because it means that the things entering his body are pure. If he seems to see the opposite of this, it is not good because it indicates that some element of disease has entered his body. Such a case should be treated as the one described above.

If it seems to rain with gentle rain from clear skies, and without any violent downpour or heavy storm, it is good. Such indicates that the breath drawn from the air is proportionate and pure. If the reverse happens, violent rain, storm and tempest, and the rain is not clear, it indicates the onset of disease from the respired air. A similar régime should be prescribed for this sort of case and very little food should be taken.

From the information which comes from this knowledge of the heavenly bodies, one must take precautions and follow the prescribed regimens. Pray to the gods: when the signs are good to the Sun, to Zeus of the sky, Zeus of the home, Athena of the home, to Hermes and Apollo. When the signs are the opposite, pray to the gods who avert evil, to Earth and to the Heroes, that all ills may be turned aside.

90. The following are some of the signs that foretell health: to see clearly and to hear distinctly things on the earth, to walk safely and to run safely and swiftly without fear, to see the earth smooth and well tilled and trees flourishing, laden with fruit and well-kept; to see rivers flowing normally with water clear and neither in flood nor with their flow lessened, and springs and wells similarly. All these things indicate the subject's health, and that the body, its flows, the food ingested and the excreta, are normal.

Anything seen which is the contrary, however, indicates something wrong in the body. Interference with sight or hearing indicates some malady of the head and longer early morning and after dinner walks than in the previous regimen should be ordered. If the legs are harmed, a contrary pull should be exerted by emetics and a greater indulgence in wrestling. Rough land indicates impurity in the flesh; longer walks after exercise should be ordered.

Trees that do not bear fruit indicate destruction of the human semen; if the trees are losing their leaves the cause of the trouble is wet and cold; if they are flourishing but barren, heat and dryness. In the one case, the regimen should aim at warming and drying; in the other, at cooling and moistening.

Abnormality in rivers relates to the flow of the blood. If the flow of a river be greater than usual, a superfluity of blood; if it be less, a deficiency. The regimen should aim at a decrease or an increase respectively. If the water is cloudy, some disturbance is indicated. This can be remedied by runs on a track or by walking; increased breathing disperses it.

Springs and wells relate to the bladder and in these cases diuretics should be employed.

A rough sea indicates disease of the bowels. Light and gentle laxatives should be used to effect a thorough purgation.

An earth-tremor or the shaking of a house predicts the onset of sickness when it is observed by a healthy man; a change and the restoration of health for a sick one. In the healthy, it is wise to change the regimen because it is the existing régime which is disturbing the whole body; therefore first give an emetic so that, after this, he may be fed up again gradually. But in the case of a sick man, because the body itself is undergoing a change, the same regimen should be continued.

To see land flooded with water or by the sea is a sign of illness, indicating excess fluid in the body. Prescribe emetics, fasting, exercise and a dry diet increasing little by little. Nor is it good to see the earth looking black or scorched; this shows excessive dehydration of the body and there is the risk of severe or fatal illness. Stop exercise and forbid all dry, pungent and diuretic food. Prescribe boiled barley-water and a small quantity of light food together with plenty of watery white wine to drink and lots of baths. The patient should not bath till he has eaten; then let him lie soft and relax, avoiding cold and sun. Pray to Earth, Hermes and the Heroes. To dream of diving into a lake, the sea or rivers is not a good sign as it too indicates an excess of moisture. It is advisable to use a dehydrating regimen and more exercise. In those suffering from fever, however, it is a good sign, indicating that the heat is being quenched by moisture.

91. It is a good sign for health to see anything normal about one's clothing, the size being neither too large nor too small but in accordance with one's own size. It is good to have white garments of one's own and the finest footwear. Anything too large or too small for one's limbs is not good; in the one case the regimen should aim at a decrease, and in the other, an increase. Black things indicate a more sickly or dangerous condition. Softening and moistening measures should be applied. New things denote a change.

92. To see the dead, clean, in white clothes, is good; while to receive something clean from them denotes health both of the body and the things which enter it. This is because the dead are a source of nourishment, increase and propagation, and it is a sign of health that what enters the body should be clean. On the contrary, if the dead appear naked, or in dark garments, or unclean, or taking or carrying anything out of the house, this is an inexpedient sign indicating disease because the things entering the body are harmful. These should be purged away by circular runs and walks and, after an emetic, a soft and light diet should be given which is gradually increased.

93. The appearance of monstrous creatures which appear during sleep and frighten the dreamer indicate a surfeit or unaccustomed food, a secretion, cholera and a dangerous illness. An emetic should be followed by an increasing diet of the lightest foods for five days; the food should neither be excessive nor pungent, nor dry nor warm. Prescribe also exercise, especially natural exercise, but not walks after dinner. Warm baths and relaxation are also advisable, and both the sun and the cold should be avoided.

To seem, while sleeping, to eat or to drink one's normal diet indicates undernourishment and a mental hunger. The stronger the meats seem, the greater the degree of inadequacy of the diet; weaker meats indicate a smaller deficiency, as if it were good to partake of whatever were seen in the dream. . . .* The diet should therefore be reduced, as it indicates a surfeit of nourishment. To dream of loaves made with cheese and honey has a similar significance.

Drinking clear water is not harmful; all other sorts of water are. Any normal things seen in a dream indicate a similar appetite of the soul.

If the dreamer flies in fright from anything, this means an obstruction to the blood as a result of dehydration. It is then wise to cool and moisten the body.

Fighting, being stabbed or bound by another indicates that some secretion, inimical to the flows, has taken place into the body. It is then advisable to take an emetic, to go on a reducing diet and to go for walks. A light diet, increasing over four days, should be taken after the emetic. Wandering and difficult climbs have the same meaning.

Fording rivers, enemy soldiers and monstrous apparitions denote illness or madness. After an emetic give a small diet of light soft food, increasing gently for five days, together with plenty of natural exercise – except walks after dinner – warm baths and relaxation. Avoid cold and sun.

By following the instructions I have given, one may live a healthy life; and I have discovered the best regimen that can be devised by a mere mortal with the help of the gods.

*A sentence is probably lost here.

THE NATURE OF MAN

A popular lecture on physiology

1. This lecture is not intended for those who are accustomed to hear discourses which inquire more deeply into the human constitution than is profitable for medical study. I am not going to assert that man is all air, or fire, or water, or earth, or in fact anything but what manifestly composes his body; let those who like discuss such matters. Nevertheless, when these things are discussed I perceive a certain discrepancy in the analyses for, although the same theory is employed, the conclusions do not agree. They all, theorizing, draw the same deduction, asserting that there is one basic substance which is unique and the basis of everything; but they call it by different names, one insisting that it is air, another that it is fire, another water, another earth. Each adds arguments and proofs to support his contention, all of which mean nothing. Now, whenever people arguing on the same theory do not reach the same conclusion, you may be sure that they do not know what they are talking about. A good illustration of this is provided by attending their disputations when the same disputants are present and the same audience; the same man never wins the argument three times running, it is first one and then the other and sometimes the one who happens to have the glibbest tongue. Yet it would be expected that the man who asserts that he can provide the correct explanation of the subject, if, that is, he really knows what he is talking about and demonstrates it correctly, should always win the argument. I am of the opinion that these people wreck their own theories on the problem of the terms they use for the One because they fail to understand the issue. Thus they serve, rather, to establish the theory of Melissus.*

*Flourished about 440 B.C.; like Parmenides, he denied plurality and change and held that what is is one and unchanging.

2. I need say no more about these theorists. But when we come to physicians, we find that some assert that man is composed of blood, others of bile and some of phlegm. But these, too, all make the same point, asserting that there is a basic unity of substance, although they each give it a different name and so change its appearance and properties under stress of heat and cold, becoming sweet or bitter, white or black, and so forth. Now I do not agree with these people either, although the majority will declare that this, or something very similar, is the case. I hold that if man were basically of one substance, he would never feel pain, since, being one, there would be nothing to hurt. Moreover, if he should feel pain, the remedy likewise would have to be single. But in fact there are many remedies because there are many things in the body which when abnormally heated, cooled, dried or moistened by interaction, engender disease. As a result, disease has a plurality of forms and a plurality of cures.

I challenge the man who asserts that blood is the sole constituent of the human body, to show, not that it undergoes changes into all sorts of forms, but that there is a time of year or of human life when blood is obviously the sole constituent of the body. It is reasonable to suppose, were this theory true, that there is one period at which it appears in its proper form. The same applies to those who make the body of phlegm or bile.

I propose to show that the substances I believe compose the body are, both nominally and essentially, always the same and unchanging; in youth as well as in age, in cold weather as well as in warm. I shall produce proofs and demonstrate the causes both of the growth and decline of each of the constituents of the body.

3. In the first place, generation cannot arise from a single substance. For how could one thing generate another unless it copulated with some other? Secondly, unless the things which copulated were of the same species and had the same generative capabilities, we should not get these results. Again, generation would be impossible unless the hot stood in a fair and reasonable proportion to the cold, and likewise the dry to

the wet; if, for instance, one preponderated over the other, one being much stronger and the other much weaker. Is it likely, then, that anything should be generated from one thing, seeing that not even a number of things suffice unless they are combined in the right proportions? It follows, then, such being the nature of the human body and of everything else, that man is not a unity but each of the elements contributing to his formation preserves in the body the power which it contributed. It also follows that each of the elements must return to its original nature when the body dies; the wet to the wet, the dry to the dry, the hot to the hot and the cold to the cold. The constitution of animals is similar and of everything else too. All things have a similar generation and a similar dissolution, for all are formed of the substances mentioned and are finally resolved in the same constituents as produced them; that too is how they disappear.

4. The human body contains blood, phlegm, yellow bile and black bile. These are the things that make up its constitution and cause its pains and health. Health is primarily that state in which these constituent substances are in the correct proportion to each other, both in strength and quantity, and are well mixed. Pain occurs when one of the substances presents either a deficiency or an excess, or is separated in the body and not mixed with the others. It is inevitable that when one of these is separated from the rest and stands by itself, not only the part from which it has come, but also that where it collects and is present in excess, should become diseased, and because it contains too much of the particular substance, cause pain and distress. Whenever there is more than slight discharge of one of these humours outside the body, then its loss is accompanied by pain. If, however, the loss, change or separation from the other humours is internal, then it inevitably causes twice as much pain, as I have said, for pain is produced both in the part whence it is derived and in the part where it accumulates.

5. Now I said that I would demonstrate that my proposed constituents of the human body were always constant, both nominally and essentially. I hold that these constituents are

blood, phlegm and yellow and black bile. Common usage has assigned to them specific and different names because there are essential differences in their appearance. Phlegm is not like blood, nor is blood like bile, nor bile like phlegm. Indeed, how could they be alike when there is no similarity in appearance and when they are different to the sense of touch? They are dissimilar in their qualities of heat, cold, dryness and moisture. It follows then that substances so unlike in appearance and characteristics cannot basically be identical, at least if fire and water are not identical. As evidence of the fact that they are dissimilar, each possessing its own qualities and nature, consider the following case. If you give a man medicine which brings up phlegm, you will find his vomit is phlegm; if you give him one which brings up bile, he will vomit bile. Similarly, black bile can be eliminated by administering a medicine which brings it up, or, if you cut the body so as to form an open wound, it bleeds. These things will take place just the same every day and every night, winter and summer, so long as the subject can draw breath and expel it again, or until he is deprived of any of these congenital elements. For they must be congenital, firstly because it is obvious that they are present at every age so long as life is present and, secondly, because they were procreated by a human being who had them all and mothered in a human being similarly endowed with all the elements which I have indicated and demonstrated.

6. Those who assert that the human body is a single substance seem to have reasoned along the following lines. Having observed that when men died from excessive purgation following the administration of drugs, some vomited bile and some phlegm, they concluded from this that whatever was the nature of the material voided at death, this was indeed the fundamental constituent of man. Those who insist that blood is the basic substance use a similar argument; because they see blood flowing from the body in the fatally wounded, they conclude that blood constitutes the soul. They all use similar arguments to support their theories. But, to begin with, no one ever yet died from excessive purgation and brought up only bile; taking medicine which causes the bringing up

of bile, produces first the vomiting of bile, but subsequently, the vomiting of phlegm as well. This is followed by the vomiting of black bile in spite of themselves and they end up by vomiting pure blood and that is how they die. The same effects result from taking a drug which brings up phlegm; the vomiting of phlegm is followed by yellow bile, then black bile, then pure blood, and so death ensues. When a drug is ingested, it first causes the evacuation of whatever in the body is naturally suited to it, but afterwards, it causes the voiding of other substances too. It is similar in the case of plants and seeds; when these are put into the ground, they first absorb the things which naturally suit them; they may be acid, bitter, sweet, salty and so forth. But although at first the plant takes what is naturally suited to it, afterwards it absorbs other things as well. The action of drugs in the body is similar; those which cause the bringing up of bile at first bring it up undiluted, but later on it is voided mixed with other substances; the same is true of drugs which bring up phlegm. In the case of men who have been fatally wounded the blood at first runs very warm and red, but subsequently it becomes more like phlegm and bile.

7. Now the quantity of phlegm in the body increases in winter because it is that bodily substance most in keeping with the winter, seeing that it is the coldest. You can verify its coldness by touching phlegm, bile and blood; you will find that the phlegm is the coldest. It is however the most viscous and is brought up with greater force than any other substance with the exception of black bile. Although those things which are forcibly expelled become warmer owing to the force to which they are subjected, nevertheless phlegm remains the coldest substance, and obviously so, owing to its natural characteristics. The following signs show that winter fills the body with phlegm: people spit and blow from their noses the most phlegmatic mucus in winter; swellings become white especially at that season and other diseases show phlegmatic signs.

During the spring, although the phlegm remains strong in the body, the quantity of blood increases. Then, as the cold becomes less intense and the rainy season comes on, the wet and warm days increase further the quantity of blood. This part of the year is most in keeping with blood because it is wet and hot. That this is so, you can judge by these signs: it is in spring and summer that people are particularly liable to dysentery and to epistaxis, and these are the seasons too at which people are warmest and their complexions are ruddiest.

During the summer, the blood is still strong but the bile gradually increases, and this change continues into the autumn when the blood decreases since the autumn is contrary to it. The bile rules the body during the summer and the autumn. As proof of this, it is during this season that people vomit bile spontaneously, or, if they take drugs, they void the most bilious sort of matter. It is plain too from the nature of fevers and from people's complexions in that season. During the summer, the phlegm is at its weakest since this season, on account of its dryness and heat, is most contrary to that substance.

The blood in the body reaches its lowest level in autumn, because this is a dry season and the body is already beginning to cool. Black bile is strongest and preponderates in the autumn. When winter sets in the bile is cooled and decreases while the phlegm increases again owing to the amount of rain and the length of the nights.

All these substances, then, are all always present in the body but vary in their relative quantities, each preponderating in turn according to its natural characteristics. The year has its share of all the elements: heat, cold, dryness and wetness. None of these could exist alone for a moment, while, on the other hand, were they missing, all would disappear, for they are all mutually interdependent. In the same way, if any of these primary bodily substances were absent from man, life would cease. And just as the year is governed at one time by winter, then by spring, then by summer and then by autumn; so at one time in the body phlegm preponderates, at another time blood, at another time yellow bile and this is followed by the preponderance of black bile. A very clear proof of this can be obtained by giving the same man the same emetic at four different times in the year; his vomit will be most phlegmatic in winter, most wet in spring, most bilious in summer and darkest in autumn.

8. In these circumstances it follows that the diseases which increase in winter should decrease in summer and vice versa. Those which come to an end in a given number of days are exceptions and I will discuss periodicity later on. You may expect diseases which begin in spring to end in the autumn; likewise autumnal diseases will disappear in the spring. Any disease which exceeds these limits must be put down as belonging to a whole year. In applying his remedies, the physician must bear in mind that each disease is most prominent during the season most in keeping with its nature.

9. In addition to these considerations, certain further points should be known. Diseases caused by over-eating are cured by fasting; those caused by starvation are cured by feeding up. Diseases caused by exertion are cured by rest; those caused by indolence are cured by exertion. To put it briefly: the physician should treat disease by the principle of opposition to the cause of the disease according to its form, its seasonal and age incidence, countering tenseness by relaxation and vice versa. This will bring the patient most relief and seems to me to be the principle of healing.

Some diseases are produced by the manner of life that is followed; others by the life-giving air we breathe. That there are these two types may be demonstrated in the following way. When a large number of people all catch the same disease at the same time, the cause must be ascribed to something common to all and which they all use; in other words to what they all breathe. In such a disease, it is obvious that individual bodily habits cannot be responsible because the malady attacks one after another, young and old, men and women alike, those who drink their wine neat and those who drink only water; those who eat barley-cake as well as those who live on bread, those who take a lot of exercise and those who take but little. The régime cannot therefore be responsible where people who live very different lives catch the same disease.

However, when many different diseases appear at the same time, it is plain that the regimen is responsible in individual cases. Treatment then should aim at opposing the cause of the disease as I have said elsewhere; that is, treatment should involve a change in regimen. For, in such a case, it is obvious that all, most, or at least one of the factors in the regimen does not agree with the patient; such must be sought out and changed having regard to the constitution of the patient, his age and appearance, the season of the year and the nature of the disease. The treatment prescribed should vary accordingly by lessening this or increasing that, and the regimen and drugs should be appropriately adapted to the various factors already mentioned.

When an epidemic of one particular disease is established, it is evident that it is not the regimen but the air breathed which is responsible. Plainly, the air must be harmful because of some morbid secretion which it contains. Your advice to patients at such a time should be not to alter the regimen since this is not to blame, but they should gradually reduce the quantity of food and drink taken so that the body is as little loaded and as weak as possible. A sudden change of regimen involves the risk of starting a fresh complaint, so you should deal with the regimen in this way when it is clearly not the cause of the patient's illness. Care should be taken that the amount of air breathed should be as small as possible and as unfamiliar as possible. These points may be dealt with by making the body thin so that the patient will avoid large and frequent breaths, and, wherever practicable, by a change of station from the infected area.

10. The most serious diseases are those which arise from the strongest part of the body, since if a disease remains in the place where it begins, it is inevitable that the whole body should sicken if its strongest part does. Alternatively, if the disease passes from the stronger part to a weaker part, it proves difficult to dispel. Those which pass from a weak part to a stronger are more easily cured because the in-flowing humours are easily spent by the strength of the part.

11. The blood-vessels of largest calibre, of which there are four pairs in the body, are arranged in the following way: one pair runs from the back of the head, through the neck, and,

weaving its way externally along the spine, passes into the legs, traverses the calves and the outer aspect of the ankle, and reaches the feet. Venesection for pains in the back and loins should therefore be practised in the hollow of the knee or externally at the ankle.

The second pair of blood-vessels runs from the head near the ears through the neck, where they are known as the jugular veins. Thence they continue deeply close to the spine on either side. They pass close to the muscles of the loins, entering the testicles and the thighs. Thence they traverse the popliteal fossa on the medial side and passing through the calves lie on the inner aspect of the ankles and the feet. Venesection for pain in the loin and in the testicles should therefore be done in the popliteal area or at the inner side of the ankle.

The third pair of blood-vessels runs from the temples, through the neck and under the shoulder-blades. They then come together in the lungs; the right-hand one crossing to the left, the left-hand one crossing to the right. The right-hand one proceeds from the lungs, passes under the breast and enters the spleen and the kidneys. The left-hand one proceeds to the right on leaving the lungs, passes under the breast and enters the liver and the kidneys. Both vessels terminate in the anus.

The fourth pair runs from the front of the head and the eyes, down the neck and under the clavicles. They then course on the upper surface of the arms as far as the elbows, through the forearms into the wrists and so into the fingers. They then return from the fingers running through the ball of the thumb and the forearms to the elbows where they course along the inferior surface of the arms to the axillae. Thence they pass superficially down the sides, one reaching the spleen and its fellow the liver. Thence they course over the belly and terminate in the pudendal area.

Apart from the larger vessels which are thus accounted for, there are a large number of vessels of all sizes running from the belly to all parts of the body; these carry foodstuffs to the body. They also form connections between the large main vessels which run to the belly and the rest of the body. In addition they join up with each other and form connections between the deep and superficial vessels.

The following are therefore rules for venesection. Care should be taken that the cuts are as close as possible to the determined source of the pain and the place where the blood collects. By doing this a sudden, violent change is avoided but at the same time the customary site of collection of blood will be changed.

12. If a patient over the age of thirty-five expectorates much without showing fever, passes urine exhibiting a large quantity of sediment painlessly, or suffers continuously from bloody stools as in cases of dysentery, his complaint will arise from the following single cause. He must, when a young man, have been hard-working, fond of physical exertion and work and then, on dropping the exercises, have run to soft flesh very different from that which he had before. There must be a sharp distinction between his previous and his present bodily physique so that the two do not agree. If a person so constituted contracts some disease, he escapes for the time being but, after the illness, the body wastes. Fluid matter then flows through the blood-vessels wherever the widest way offers. If it makes its way to the lower bowel it is passed in the stools in much the same form as it was in the body; as its course is downwards it does not stay long in the intestines. If it flows into the chest, suppuration results because, owing to the upward tread of its path, it spends a long time in the chest and there rots and forms pus. Should the fluid matter, however, be expelled into the bladder, it becomes warm and white owing to the warmth of that region. It becomes separated in the urine; the lighter elements float and form a scum on the surface while the heavier constituents fall to the bottom forming pus.

Children suffer from stones owing to the warmth of the whole body and of the region about the bladder in particular. Adult men do not suffer from stone because the body is cool; it should be thoroughly appreciated that a person is warmest the day he is born and coldest the day he dies. So long as the body is growing and advancing towards strength it is necessarily warm; but when it begins to wither and to fade away to feebleness, it cools down. From this principle it follows that a person is warmest the day he is born because he grows most on that day; he is coldest the day he dies because on that day he withers most.

People of the constitution mentioned above, that is athletic people who have got soft, generally recover of their own accord within forty-five days of the wasting beginning. If such a period be exceeded, natural recovery takes a year so long as no other malady intervenes.

13. Prognosis is safest to foretell in those diseases which develop quickly and those whose causes are apparent. They should be cured by opposing whatever is the cause of the disease, of which the body will thus be rid.

14. The presence of a sandy sediment or of stones in the urine means that originally tumours grew in relation to the aorta and suppurated. Then, because the tumour did not burst rapidly, stones were formed from the pus and these were squeezed out through the blood-vessels together with urine into the bladder. When the urine is only blood-stained, the blood-vessels have been attacked. Sometimes the urine is thick and small hair-like pieces of flesh are voided with it which, it must be realized, come from the kidneys and the joints. When in an otherwise clear urine, a substance like bran is present in it, the bladder is inflamed.

15. Most fevers are caused by bile. Apart from those arising from local injury, they are of four types. These are called continued, quotidian, tertian and quartan.

Continued fever is produced by large quantities of the most concentrated bile and the crisis is reached in the shortest time; as the body enjoys no periods of coolness, the great heat it endures results in rapid wasting.

Quotidian fever is caused by a large quantity of bile, but less than that which causes continued fever. This is quicker than the others to depart although it lasts longer than a continued fever by as much as there is less bile causing it, and because the body has some respite from the fever whereas continued fever allows none.

Tertian fever lasts longer than quotidian fever and is caused by less bile. A tertian fever is longer in proportion to the longer respites from fever allowed to the body compared with quotidian fevers.

Quartans behave similarly to the tertians but last longer, as they arise from still less of the heat-producing bile and because they give the body longer respites in which to cool down. A secondary reason for their chronic character and difficult resolution is that they are caused by black bile; this is the most viscous of the humours in the body and remains the longest. As evidence of this note the association of quartan fevers with melancholy. Quartan fever has its highest incidence in the autumn and in those between the ages of twenty-five and forty-five. This is the time of life when the body is most subject to black bile, and the autumn is the corresponding season of the year. If a quartan fever occurs at any other time of the year, or at any other age, you may be sure that it will not be chronic unless some other malady be present.

A REGIMEN FOR HEALTH

An early recognition of the importance of preventive medicine

1. The ordinary man should adopt the following regimen. During the winter, he should eat as much as possible, drink as little as possible and this drink should be wine as undiluted as possible. Of cereals, he should eat bread, all his meat and fish should be roasted and he should eat as few vegetables as possible during winter-time. Such a diet will keep the body warm and dry.

When spring comes, he should take more to drink, increasing the quantity and making it more watery, a little at a time. He should take softer cereals and less of them, substituting barley-cake for bread. Meat should be cut down in the same way, boiled meat replacing roast, and a few vegetables should be eaten once spring has begun. Thus will he effect a gradual change and towards the summer he will be taking a diet consisting entirely of soft cereals, boiled meat and vegetables both raw and boiled. At that time he will be taking the greatest quantity of the most diluted wine, taking care that the change is neither violent nor sudden but that it is made gradually.

During the summer he should live on soft barley-cake, watered wine in large quantities and take all his meat boiled. Such a diet is necessary in summer to make the body cool and soft, for the season, being hot and dry, renders the body burnt-up and parched, and such a condition may be avoided by a suitable diet. The change from spring to summer should follow the same pattern as that from winter to spring, decreasing the amount of cereals and increasing the quantity of drink taken.

A reversal of this process constitutes the transition from the summer to the winter diet. In the autumn the cereals should be increased and made drier, and likewise the meat in the diet. The quantity of drink taken should be decreased and taken less diluted so that he will have a good winter. Once more, then, he takes the smallest quantity of the least diluted drink and the largest quantity of cereals of the driest kind. This will keep him in good health and he will feel the cold less, for the season is cold and wet.

2. People with a fleshy, soft or ruddy appearance are best kept on a dry diet for the greater part of the year as they are constitutionally moist. Those with firm and tight-drawn skins, and those with tawny or dark complexions, should keep to a diet containing plenty of fluid most of the time, as the constitutions of such people are naturally dry. The softest and most moist diets suit young bodies best as at that age the body is dry and has set firm. Older people should take a drier diet most of the time, for at that age bodies are moist, soft and cold. Diets then must be conditioned by age, the time of year, habit, country and constitution. They should be opposite in character to the prevailing climate, whether winter or summer. Such is the best road to health.

3. In winter a man should walk quickly, in summer in more leisurely fashion unless he is walking in the hot sun. Fleshy people should walk faster, thin people more slowly. More baths should be taken in summer than in winter; firm people should bathe more than the fleshy ones. Garments in summer should be steeped in olive oil, but not in winter.

4. Fat people who want to reduce should take their exercise on an empty stomach and sit down to their food out of breath. They should not wait to recover their breath. They should before eating drink some diluted wine, not too cold, and their meat should be dished up with sesame seeds or seasoning and such-like things. The meat should also be fat as the smallest quantity of this is filling. They should take only one meal a day, go without baths, sleep on hard beds and walk about with as little clothing as may be. Thin people who want to get fat should do exactly the opposite and never take exercise on an empty stomach.

5. The following rules are to be observed in the administration of emetics and enemata. Vomiting may be induced during the six winter months, as this is the phlegmatic time of year

and diseases are centred about the head and in the chest. During the warm weather, enemata may be used as this is the hot season when the body is more bilious and heaviness occurs in the loins and knees, when there are fevers and colic in the belly. It is necessary therefore to cool the body and draw downwards the matter surrounding those regions. For those who are rather fat and moist, use the thinner and more briny enemata; for those who are drier and have firmer flesh, use the more fatty and thicker enemata. By fatty and thicker enemas, I mean those made from milk and from chick-pea and boiled water and such-like; by thin and briny, such things as brine and sea-water.

Emetics should be administered as follows. Those who are fat, but not those who are lean, should vomit on an empty stomach after a run or a brisk walk about the middle of the day. The emetic should consist of a gill of ground hyssop in six pints of water; this should be drunk after adding vinegar and salt to improve the taste. It should at first be drunk slowly, but the remainder more quickly. Thinner and weaker people should take emetics after food in the following way. A hot bath should be followed by drinking half a pint of neat wine after which a meal of any kind of food should be taken, but no drink is taken either with the meal or after it. Wait as long as it takes to walk a mile and then administer a mixture of three wines, a bitter, a sweet and an acid one, at first neat in small doses at long intervals and then more diluted in larger doses and more frequently.

Those who are accustomed to induce vomiting twice a month will find it better to do so on two consecutive days rather than every fortnight; as it is, most people do the opposite. Those who benefit from vomiting and those who have difficulty with passing stools should eat several times a day and take all varieties of food and their meat cooked in every different way and drink two or three kinds of wine. The opposite kind of diet is best for those who do not indulge in vomiting or for those with relaxed bellies.

6. Infants should be bathed for long periods in warm water and given their wine diluted and not at all cold. The wine

should be of a kind which is least likely to cause distension of the stomach and wind. This should be done to prevent the occurrence of convulsions and to make the children grow and get good complexions. Women do best on a drier diet as dry foods are most suited to the softness of their flesh, and the less diluted drinks are better for the womb and for pregnancy.

7. Those who enjoy gymnastics should run and wrestle during the winter; in summer, wrestling should be restricted and running forbidden, but long walks in the cool part of the day should take their place. Those who get exhausted with running should wrestle, and those who get exhausted with wrestling should run. By exercising in this way, the exhausted parts of the body will best be warmed, composed and rested.

Those who find that exercise causes diarrhoea and who pass undigested stools resembling food, should have their exercise cut by at least a third while their food should be halved. For it is clear that the belly cannot get sufficiently warm to digest the greater part of the food. The diet in such cases should consist of bread baked as well as possible crumbled in wine, together with the smallest quantity of practically undiluted wine. They should not walk after meals. They should also take only one meal a day during the time they have diarrhoea; this will give the belly the best chance to deal with the food that is given it. This sort of diarrhoea is most common in those who are particularly stout, when, their constitution being what it is, they are obliged to eat meat. The vessels being compressed, they cannot cope with the intake of food. This type of constitution is nicely balanced, liable to fall off in either direction and it is at its best for only a short time.

The sparer and more hirsute type of person can better cope with a big diet and also with hard exercise. They remain at the height of their powers for a longer period.

Those who vomit their food the day after it has been taken and suffer from distension of the hypochondrium showing that the food remains undigested, should take more sleep and force their bodies by exercise. They should drink more wine and take it less diluted and also, at these times, reduce the amount of food. For it is clear that the weakness and coldness of the belly prevent the greater part of the food from being digested.

Those who suffer from thirst should reduce both the amount of food and the amount of exercise they take, and they should be given watery wine to drink as cold as possible.

Those who get pains in the viscera as the result of gymnastics or any other form of exercise should rest without eating and drink the smallest quantity necessary to cause the passing of the greatest amount of urine. In this way the vessels coursing through the viscera will not become filled and distended and so cause tumours and fevers.

9.* A wise man ought to realize that health is his most valuable possession and learn how to treat his illnesses by his own judgement.

---

*Chapter 8, which is an interpolation from another work, is omitted.

SURGERY

TRANSLATED BY E. T. WITHINGTON

FRACTURES

One of the great Hippocratic surgical treatises.

1. In dislocations and fractures, the practitioner should make extensions in as straight a line as possible, for this is most conformable with nature;* but if it inclines at all to either side, it should turn towards pronation [palm down] rather than supination [palm up], for the error is less. Indeed, those who have no preconceived idea make no mistake as a rule, for the patient himself holds out the arm for bandaging in the position impressed on it by conformity with nature. The theorizing practitioners are just the ones who go wrong. In fact the treatment of a fractured arm is not difficult, and is almost any practitioner's job, but I have to write a good deal about it because I know practitioners who have got credit for wisdom by putting up arms in positions which ought rather to have given them a name for ignorance. And many other parts of this art are judged thus: for they praise what seems outlandish before they know whether it is good, rather than the customary which they already know to be good; the bizarre rather than the obvious. One must mention then those errors of practitioners as to the nature of the arm on which I want to give positive and negative instruction, for this discourse is an instruction on other bones of the body also.

2. To come to our subject, a patient presented his arm to be dressed in the attitude of pronation, but the practitioner made him hold it as the archers do when they bring forward the shoulder, and he put it up in this posture, persuading himself that this was its natural position. He adduced as evidence the parallelism of the forearm bones, and the surface also, how that it has its outer and inner parts in a direct line, declaring

*Galen makes this a general statement; but the writer is apparently speaking of the forearm, which he had already mentioned in a lost introduction.

this to be the natural disposition of the flesh and tendons, and he brought in the art of the archer as evidence. This gave an appearance of wisdom to his discourse and practice, but he had forgotten the other arts and all those things which are executed by strength or artifice, not knowing that the natural position varies in one and another, and that in doing the same work it may be that the right arm has one natural position and the left another. For there is one natural position in throwing the javelin, another in using the sling, another in casting a stone, another in boxing, another in repose. How many arts might one find in which the natural position of the arms is not the same, but they assume postures in accordance with the apparatus each man uses and the work he wants to accomplish! As to the practiser of archery, he naturally finds the above posture strongest for one arm: for the hinge-like end of the humerus in this position being pressed into the cavity of the ulna makes a straight line of the bones of the upper arm and forearm, as if the whole were one, and the flexure of the joint is extended [abolished] in this attitude. Naturally then the part is thus most inflexible and tense, so as neither to be overcome or give way when the cord is drawn by the right hand. And thus he will make the longest pull, and shoot with the greatest force and frequency, for shafts launched in this way fly strongly, swiftly and far. But there is nothing in common between putting up fractures and archery. For, first, if the operator, after putting up an arm, kept it in this position, he would inflict much additional pain, greater than that of the injury, and again, if he bade him bend the elbow, neither bones, tendons, nor flesh would keep in the same position, but would rearrange themselves in spite of the dressings. Where, then, is the advantage of the archer position? And perhaps our theorizer would not have committed this error had he let the patient himself present the arm.

3. Again, another practitioner handing over the arm back downwards had it extended thus and then put it up in this position, supposing it to be the natural one from surface indications: presuming also that the bones are in their natural position because the prominent bone at the wrist on the little

finger side appears to be in line with the bone from which men measure the forearm [cubit]. He adduced this as evidence for the naturalness of the position, and seemed to speak well.

But, to begin with, if the arm were kept extended in supination it would be very painful; anyone who held his arm extended in this position would find how painful it is. In fact, a weaker person grasping a stronger one firmly so as to get his elbow extended in supination might lead him whither he chose, for if he had a sword in this hand he would be unable to use it, so constrained is this attitude. Further, if one put up a patient's arm in this position and left him so, the pain, though greater when he walked about, would also be great when he was recumbent. Again, if he shall bend the arm, it is absolutely necessary for both the muscles and bones to have another position. Besides the harm done, the practitioner was ignorant of the following facts as to the position. The projecting bone at the wrist on the side of the little finger belongs indeed to the ulna, but that at the bend of the elbow from which men measure the cubit is the head of the humerus, whereas he thought the one and the other belonged to the same bone, and so do many besides. It is the so-called elbow on which we lean that belongs to this bone.* In a patient with the forearm thus supinated, first, the bone is obviously distorted, and secondly, the cords stretching from the wrist on its inner side and from the fingers also undergo distortion in this supine position, for these cords extend to the bone of the upper arm from which the cubit is measured. Such and so great are these errors and ignorances concerning the nature of the arm. But if one does extension of a fractured arm as I direct, he will both turn the bone stretching from the region of the little finger to the elbow so as to be straight† and will have the cords stretching from the wrist to the [lower] end of the humerus in a direct line; further, the arm when slung will keep about the same position as it was in when put up, and it will give the patient no pain when he walks, no pain when he lies down and no sense of weariness. The patient should be so seated that the projecting

*i.e. the olecranon process is part of the ulna.
†i.e. the styloid process in line with the olecranon.

part of the bone is turned towards the brightest light available, that the operator may not overlook the proper degree of extension and straightening. Of course the hand of an experienced practitioner would not fail to recognize the prominence [at the fracture] by touch; also there is a special tenderness at the prominence when palpated.

4. When the bones of the forearm are not both fractured the cure is easier if the upper bone [radius] is injured, though it is the thicker, both because the sound bone lying underneath acts as a support and because it is better covered, except at the part near the wrist, for the fleshy growth on the upper bone is thick; but the lower bone [ulna] is fleshless, not well covered, and requires stronger extension. If it is not this bone but the other that is broken, rather slight extension suffices: if both are broken very strong extension is requisite. In the case of a child I have seen the bones extended more than was necessary, but most patients get less than the proper amount. During extension one should use the palms of the hands to press the parts into position, then after anointing with cerate (in no great quantity lest the dressings should slip), proceed to put it up in such a way that the patient shall have his hand not lower than the elbow but a little higher; so that the blood may not flow to the extremity but be kept back. Then apply the linen bandage, putting the head of it at the fracture so as to give support, but without much pressure. After two or three turns are made at the same spot, let the bandage be carried upwards that afflux of blood may be kept back, and let it end off there. The first bandages should not be lengthy. Put the head of the second bandage on the fracture, making one turn there; then let it be carried downwards, with decreasing pressure and at wider intervals, till enough of the bandage is left for it to run back again to the place where the other ended. Let the bandages in this part of the dressing be applied either to left or right, whichever suits the form of the fracture and the direction towards which the limb ought to turn. After this, compresses should be laid along after being anointed with a little cerate; for the application is more supple and more easily made. Then put on bandages crosswise to right and

left alternately, beginning in most cases from below upwards but sometimes from above downwards. Treat conical parts by surrounding them with compresses, bringing them to a level not all at once but gradually by the number of circumvolutions. You should put additional loose turns now and then at the wrist. The two sets of bandages are a sufficient number for the first dressing.

5. These are the indications of good treatment and correct bandaging: If you ask the patient whether the part is compressed and he says it is, but moderately and that chiefly at the fracture. A properly bandaged patient should give a similar report of the operation throughout. The following are the indications of a due moderation. During the day of the dressing and the following night the pressure should appear to the patient not to diminish but rather to increase, and on the following day a slight and soft swelling should appear in the hand; you should take this as a sign of the due mean as to pressure. At the end of the day the pressure should seem less, and on the third day you should find the bandages loose. If, then, any of the said conditions are lacking you may conclude that the bandaging was slacker than the mean, but if any of them be excessive you may conclude that the pressure was greater than the mean, and taking this as a guide make the next dressing looser or tighter. You should remove the dressing on the third day after the extension and adjustment, and if your first bandaging hit the proper mean this one should be a little tighter. The heads of the bandages should be applied over the fracture as before, for if you did this before, the serous effusions were driven thence into the outer parts on both sides, but if you formerly made the pressure anywhere else, they were driven into this place [the fracture] from the part compressed. It is useful for many things to understand this. It shows that one should always begin the bandaging and compression at this point, and, for the rest, in proportion as you get further from the point of fracture make the pressure less. Never make the turns altogether slack, but closely adherent. Further, one should use more bandages at each dressing, and the patient when asked should say he felt a little more pressure

than before, especially at the point of fracture, and the rest in proportion. And as regards the swelling, feeling of pain and relief, things should be in accord with the previous dressing. When the third day comes, he should find the dressings rather loose. Then after undoing them he should bandage again with a little more pressure and with all the bandages that he is going to use, and afterwards the patient should experience all those symptoms which he had in the first periods of bandaging.

6. When the third day is reached (the seventh from the first dressing), if he is being properly bandaged, there will be the swelling on the hand, but it will not be very marked. As to the part bandaged, it will be found to be thinner and more shrunken at each dressing, and on the seventh day it will be quite thin, while the fractured bones will be more mobile and ready for adjustment. If this is so, after seeing to the adjustment you should bandage as for splints, making a little more pressure than before, unless there is any increase of pain from the swelling on the hand. When you dress with the bandages you should apply the splints round the limb and include them in ligatures as loose as possible consistently with firmness, so that the addition of the splints may contribute nothing to the compression of the arm. After this the pain and the relief following it should be the same as in the previous periods of bandaging. When, on the third day, he says it is loose, then indeed you should tighten up the splints, especially at the fracture, and the rest in proportion where the dressing also was loose rather than tight. The splint should be thicker where the fracture projects, but not much so, and you should take special care that it does not lie in the line of the thumb, but on one side or the other, nor in the line of the little finger where the bone projects at the wrist, but on one side or the other. If, indeed, it is for the benefit of the fracture that some of the splints should be placed thus, you should make them shorter than the rest, so that they do not reach as far as the bones which project at the wrist, for there is risk of ulceration and denuding of tendons. You should tighten the splints every third day very slightly, bearing in mind that they are put there

to maintain the dressing, but not bound in for the sake of pressure.

7. If you are convinced that the bones are sufficiently adjusted in the former dressings, and there is no painful irritation nor any suspicion of a sore, you should leave the part put up in splints till over the twentieth day. It takes about thirty days altogether as a rule for the bone of the forearm to unite. But there is nothing exact about it, for both constitutions and ages differ greatly. When you remove the dressing, douche with warm water and replace it, using a little less pressure and fewer bandages than before; and after this, remove and re-apply every other day with less pressure and fewer bandages. If, in any case where splints are used, you suspect that the bones are not properly adjusted, or that something else is troubling the patient, remove the dressing and replace it in the middle of the interval or a little sooner. Light diet suffices in those cases where there is no open wound at the first, or protrusion of the bone, for it should be slightly restricted for the first ten days, seeing that the patients are resting; and soft foods should be taken such as favour a due amount of evacuation. Avoid wine and meat, but afterwards gradually feed him up. This discourse gives a sort of normal rule for the treatment of fractures, how one should handle them surgically, and the results of correct handling. If any of the results are not as described, you may be sure there has been some defect or excess in the surgical treatment. You should acquaint yourself further with the following points in this simple method, points with which practitioners do not trouble themselves very much, though they are such as (if not properly seen to) can bring to naught all your carefulness in bandaging. If both bones are broken, or the lower [ulna] only, and the patient, after bandaging, has his arm slung in a sort of scarf, this scarf being chiefly at the point of fracture, while the arm on either side is unsupported, he will necessarily be found to have the bone distorted towards the upper side; while if, when the bones are thus broken, he has the hand and part near the elbow in the scarf, while the rest of the arm is unsupported, this patient will be

found to have the bone distorted towards the lower side. It follows that as much as possible of the arm and wrist should be supported evenly in a soft broad scarf.

8. When the humerus is fractured, if one extends the whole arm and keeps it in this posture, the muscle of the arm* will be bandaged in a state of extension, but when the bandaged patient bends his arm the muscle will assume another posture. It follows that the most correct mode of extension of the arm is this: One should hang up a rod, in shape like a spade handle and of a cubit in length or rather shorter, by a cord at each end. Seat the patient on a high stool and pass his arm over the rod so that it comes evenly under the armpit in such a position that the man can hardly sit and is almost suspended. Then placing another stool, put one or more leather cushions under the forearm as may suit its elevation when flexed at a right angle. The best plan is to pass some broad soft leather or a broad scarf round the arm and suspend from it heavy weights sufficient for due extension; failing this, let a strong man grasp the arm in this position at the elbow and force it downwards. As to the surgeon, he should operate standing with one foot on some elevated support, adjusting the bone with the palms of his hands. The adjustment will be easy, for there is good extension if it is properly managed. Then let him do the bandaging, putting the heads of the bandages on the fracture and performing all the rest of the operation as previously directed. Let him ask the same questions, and use the same indications to judge whether things are right or not. He should bandage every third day and use greater pressure, and on the seventh or ninth day put it up in splints. If he suspects the bone is not in good position, let him loosen the dressings towards the middle of this period† and after putting it right re-apply them.

The bone of the upper arm usually consolidates in forty days. When these are passed one should undo the dressings and diminish the pressure and the number of bandages. A somewhat stricter diet and more prolonged [is required here]

---

*Biceps.   †i.e. the period in splints.

than in the former case. Make your estimate from the swelling in the hand, having an eye to the patient's strength. One must also bear in mind that the humerus is naturally convex outwards, and is therefore apt to get distorted in this direction when improperly treated. In fact, all bones when fractured tend to become distorted during the cure towards the side to which they are naturally bent. So, if you suspect anything of this kind, you should pass round it an additional broad band, binding it to the chest, and when the patient goes to bed, put a many-folded compress, or something of the kind, between the elbow and the ribs, thus the curvature of the bone will be rectified. You must take care, however, that it is not bent too much inwards.

9. The human foot, like the hand, is composed of many small bones. These bones are not often broken, unless the tissues are also wounded by something sharp or heavy. The proper treatment of the wounded parts will be discussed in the section on lesions of soft parts.* But if any of the bones be displaced, whether a joint of the toes or some bone of what is called the tarsus, you should press each back into its proper place just in the way described as regards the bones of the hand.† Treat as in cases of fracture with cerate, compresses and bandages, but without splints, using pressure in the same way and changing the dressings every other day. The patient's answers both as to pressure and relaxation should be similar to those in cases of fracture. All these bones are completely healed in twenty days, except those which are connected with the leg-bones in a vertical line. It is good to lie up during this period, but patients, despising the injury, do not bring themselves to this, but go about before they are well. This is the reason why most of them do not make a complete recovery, and the pain often returns; naturally so, for the feet carry the whole weight. It follows that when they walk about before they are well, the displaced joints heal up badly; on which account they have occasional pains in the parts near the leg.

10. [Displacement of the astragalus?] The bones which are

* Rather 'compound fractures', cf. chs. 24f.
† A lost chapter, condensed in *Joints*, ch. 26.

in connection with those of the leg are larger than the others,* and when they are displaced healing takes much longer. Treatment, indeed, is the same, but more bandages and pads should be used, also extend the dressings completely in both directions. Use pressure, as in all cases so here especially, at the point of displacement, and make the first turns of the bandage there. At each change of dressing use plenty of warm water; indeed, douche copiously with warm water in all injuries of joints. There should be the same signs as to pressure and slackness in the same periods as in the former cases, and the change of dressings should be made in the same way. These patients recover completely in about forty days, if they bring themselves to lie up; failing this, they suffer the same as the former cases, and to a greater degree.

11. Those who, in leaping from a height, come down violently on the heel, get the bones separated, while there is extravasation from the blood-vessels since the flesh is contused about the bone. Swelling supervenes and severe pain, for this bone is not small, it extends beyond the line of the leg, and is connected with important vessels and cords. The back tendon† is inserted into this bone. You should treat these patients with cerate, pads and bandages, using an abundance of hot water, and they require plenty of bandages, the best and softest you can get. If the skin about the heel is naturally smooth, leave it alone, but if thick and hard as it is in some persons, you should pare it evenly and thin it down without going through to the flesh. It is not every man's job to bandage such cases properly, for if one applies the bandage, as is done in other lesions at the ankle, taking one turn round the foot and the next round the back tendon, the bandage compresses the part and excludes the heel where the contusion is, so that there is risk of necrosis of the heel-bone; and if there is necrosis the malady may last the patient's whole life. In fact, necrosis from other causes, as when the heel blackens while the patient is in bed owing to carelessness as to its position, or when there is a serious and chronic wound in the leg connected with the heel, or in the thigh, or another malady involving prolonged rest

* Those of the wrist.   † *Tendo Achillis.*

on his back – all these necroses are equally chronic and trouble-some, and often break out afresh if not treated with most skilful attention and long rest. Necroses of this sort, indeed, besides other harm, bring great dangers to the body, for there may be very acute fevers, continuous and attended by trem-blings, hiccoughs and affections of the mind, fatal in a few days. There may be also be lividity and congestion of the large blood-vessels, loss of sensation and gangrene due to compres-sion, and these may occur without necrosis of the bone. The above remarks apply to very severe contusions, but the parts are often moderately contused and require no very great care, though, all the same, they must be treated properly. When, however, the crushing seems violent the above directions should be observed, the greater part of the bandaging being about the heel, taking turns sometimes round the end of the foot, sometimes about the middle part, and sometimes carry-ing it up the leg. All the neighbouring parts in both directions should be included in the bandage, as explained above; and do not make strong pressure, but use many bandages. It is also good to give a dose of hellebore on the first and second days. Remove the bandage and re-apply it on the third day. The following are signs of the presence and absence of aggravations. When there are extravasations from the blood-vessels, and blackenings, and the neighbouring parts become reddish and rather hard, there is danger of aggravation. Still, if there is no fever you should give an emetic as was directed; also in cases where the fever is not continuous; but if there is continued fever, do not give an evacuant, but avoid food, solid or fluid, and for drink use water and not wine, but hydro-mel may be taken. If there is not going to be aggravation, the effusions and blackenings and the parts around become yel-lowish and not hard. This is good evidence in all extravasa-tions that they are not going to get worse, but in those which turn livid and hard there is danger of gangrene. One must see that the foot is, as a rule, a little higher that the rest of the body. The patient will recover in sixty days if he keeps at rest.

12. The leg has two bones, one much more slender than the other at one end, but not so much at the other end. The parts

near the foot are joined together and have a common epiphysis. In the length of the leg they are not united, but the parts near the thigh bone are united and have an epiphysis, and the epiphysis has a diaphysis.* The bone on the side of the little toe is slightly the longer. This is the disposition of the leg-bones.

13. The bones are occasionally dislocated at the foot end, sometimes both bones with the epiphysis, sometimes the epiphysis is displaced, sometimes one of the bones. These dislocations give less trouble than those of the wrist, if the patients can bring themselves to lie up. The treatment is similar to that of the latter, for reduction is to be made by extension as in those cases, but stronger extension is requisite since the body is stronger in this part. As a rule two men suffice, one pulling one way and one the other, but if they cannot do it, it is easy to make the extension more powerful. Thus, one should fix a wheel-nave or something similar in the ground, put a soft wrapping round the foot, and then binding broad straps of ox-hide about it attach the ends of the straps to a pestle or some other rod. Put the end of the rod into the wheel-nave and pull back, while assistants hold the patient on the upper side grasping both at the shoulders and hollow of the knee. The upper part of the body can also be fixed by an apparatus. First, then, you may fix a smooth, round rod deeply in the ground with its upper part projecting between the legs at the fork, so as to prevent the body from giving way when they make extension at the foot. Also it should not incline towards the leg which is being extended, but an assistant seated at the side should press back the hip so·that the body is not drawn sideways. Again, if you like, the pegs may be fixed at either armpit, and the arms kept extended along the sides. Let someone also take hold at the knee, and so counter-extension may be made. Again, if one thinks fit, one may like-wise fasten straps about the knee and thigh, and fixing another wheel-nave in the ground above the head, attach the straps to a rod; use the nave as a fulcrum for the rod and make extension counter to that at the feet. Further, if you like, instead of the

*Spinous process or medial projection.

wheel-naves, stretch a plank of suitable length under the bed, then, using the head of the plank at each end as fulcrum, draw back the rods and make extension on the straps. And if you choose, set up windlasses at either end and make the extension by them. There are also many other methods for extensions. The best thing for anyone who practises in a large city is to get a wooden apparatus comprising all the mechanical methods for all fractures and for reduction of all joints by extension and leverage. This wooden apparatus will suffice if it be like the quadrangular supports such as are made of oak in length, breadth and thickness.

When you make sufficient extension it is then easy to reduce the joint for it is elevated in a direct line above its old position. It should therefore be adjusted with the palms of the hands, pressing upon the projecting part with one palm and with the other making counter pressure below the ankle on the opposite side.

14. After reduction, you should if possible apply a bandage, while the limb is kept extended. If the straps get in the way, remove them and keep up counter extension while bandaging. Bandage in the same way [as for fractures] putting the heads of the bandages on the projecting part and making the first and most turns there, also most of the compresses should be there and the pressure should come especially on this part. Also extend the dressing considerably to either side. This joint requires somewhat greater pressure at the first bandaging than does the wrist. After dressing let the bandaged part be higher than the rest of the body, and put it up in a position in which the foot is as little as possible unsupported.* The patient should undergo a reducing process corresponding to his strength and to the displacement, for the displacement may be small or great. As a rule the reducing treatment should be stricter and more prolonged in injuries about the leg region than in those about the arm region, for the former parts are larger and stouter than the latter. And it is especially needful for the body to be at rest and lie up. As to rebandaging the

*Not merely prevented from hanging down, but kept at right angles to the leg.

joint on the third day, there is neither hindrance nor urgency, and one should conduct all the other treatment as in the previous cases. If the patient brings himself to keep at rest and lie up, forty days are sufficient, provided only that the bones are back again in their places. If he will not keep at rest, he will not easily recover the use of the leg and will have to use bandages for a long time. Whenever the bones are not completely replaced but there is something wanting, the hip, thigh and leg gradually become atrophied. If the dislocation is inwards the outer part is atrophied, if outwards, the inner: now most dislocations are inwards.*

15. When both leg-bones are broken without an external wound, stronger extension is required. If there is much overlapping make extension by some of those methods which have been described. But extensions made by man-power are also sufficient, for in most cases two strong men are enough, one pulling at each end. The traction should be in a straight line in accordance with the natural direction of the leg and thigh, both when it is being made for fractures of the leg bones and of the thigh. Apply the bandage while both† are extended, whichever of the two you are dressing, for the same treatment does not suit both leg and arm. For when fractures of the forearm and upper arm are bandaged, the arm is slung, and if you bandage it when extended the positions of the fleshy parts are altered by bending the elbow. Further, the elbow cannot be kept extended a long time, since it is not used to that posture, but to that of flexion. And besides, since patients are able to go about after injuries of the arm, they want it flexed at the elbow. But the leg both in walking and standing is accustomed to be sometimes extended and sometimes nearly so, and it is naturally directed downwards and, what is more, its function is to support the body. Extension therefore is easily borne when necessary and indeed it frequently has this position in bed. If then it is injured, necessity brings the mind into subjection, because patients are unable to rise, so that they do not even think of bending their legs and getting up, but keep lying at

*i.e. of the foot outwards and the leg inwards.
†i.e. thigh and leg.

rest in this posture. For these reasons, then, the same position either in making extension or bandaging is unsuitable for both arm and leg. If, then, extension by man-power is enough, one should not take useless trouble, for to have recourse to machines when not required is rather absurd. But if extension by man-power is not enough, bring in some of the mechanical aids, whichever may be useful. When once sufficient extension is made, it becomes fairly easy to adjust the bones to their natural position by straightening them and making coaptation with the palms of the hands.

16. After adjustment, apply the bandages while the limb is extended, making the turns with the first bandage, either to right or left as may be suitable. Put the head of the bandage at the fracture and make the first turns there, and then carry the bandaging to the upper part of the leg as was directed for the other fractures. The bandages should be broader and longer and much more numerous for the leg parts than those of the arm. On completing the dressing, put up the limb on something smooth and soft so that it does not get distorted to either side or become concave or convex. The most suitable thing to put under is a pillow of linen or wool, not hard, making a median longitudinal depression in it, or something that resembles this.

As for the hollow splints which are put under fractured legs I am at a loss what to advise as regards their use. For the good they do is not so great as those who use them suppose. The hollow splints do not compel immobility as they think, for neither does the hollow splint forcibly prevent the limb from following the body when turned to either side, unless the patient himself sees to it, nor does it hinder the leg itself apart from the body from moving this way or that. Besides, it is, of course, rather unpleasant to have wood under the limb unless at the same time one inserts something soft. But it is very useful in changing the bedclothes, and in getting up to go to stool. It is thus possible either with or without the hollow splint to arrange the matter well or clumsily. Still the vulgar have greater faith in it, and the practitioner will be more free from blame if a hollow splint is applied, though it is rather bad practice. Anyhow, the limb should be on something smooth and soft and be absolutely straight, since it necessarily follows that the bandaging is overcome by any deviation in posture, whatever the direction or extent of it may be. The patient should give the same answers as those above mentioned, for the bandaging should be similar, and there should be the like swelling on the extremities, and so with the looseness and the changes of dressing every third day. So, too, the bandaged part should be found more slender and greater pressure be used in the dresssings and more bandages. You should also make some slack turns round the foot if the injury is not very near the knee. One should make moderate extension and adjustment of the bones at each dressing; for if the treatment be correct and the oedema subsides regularly, the bandaged part will be more slender and attenuated while the bones on their side will be more mobile and lend themselves more readily to extension. On the seventh, ninth, or eleventh day splints should be applied as was directed in the case of other fractures, and one must be careful as to the position of the splints, both in the line of the ankles, and about the back tendon from leg to foot. The bones of the leg solidify in forty days if properly treated. If you suspect that one of the bones requires some adjustment, or are afraid of ulceration, you should unbandage the part in the interval and reapply after putting it right.

17. If one of the leg-bones be broken, the extension required is weaker: there should, however, be no shortcoming or feebleness about it. Especially at the first dressing sufficient extension should be made in all fractures so as to bring the bones together, or, failing this, as soon as possible, for when one in bandaging uses pressure, if the bones have not been properly set, the part becomes more painful. The rest of the treatment is the same.

18. Of the bones, the inner of the so-called shin is the more troublesome to treat, requiring greater extension, and if the fragments are not properly set, it cannot be hid, for it is visible and entirely without flesh. When this bone is broken, patients take longer before they can use the leg, while if the

outer bone be fractured they have much less inconvenience to bear, and, even if not well set, it is much more readily concealed; for it is well covered: and they can soon stand. For the inner shin-bone carries the greatest part of the weight, since both by the disposition of the leg itself and by the direct line of the weight upon the leg the inner bone has most of the work. Further, the head of the thigh-bone sustains the body from below and has its natural direction towards the inner side of the leg and not the outer, but is in the line of the shin bone. So, too, the corresponding half of the body is nearer the line of this bone than that of the outer one, and besides, the inner is thicker than the outer, just as in the forearm the bone on the side of the little finger is longer and more slender; but in this lower articulation the longer bone does not lie underneath in the same way, for flexion at the elbow and knee are dissimilar. For these reasons, when the outer bone is fractured patients soon get about; but when the inner one is broken they do so slowly.

19. If the thigh-bone is fractured, it is most important that there should be no deficiency in the extension that is made, while any excess will do no harm. In fact, even if one should bandage while the bones were separated by the force of the extension, the dressing would have no power to keep them apart, but they would come together immediately when the assistants relaxed their tension. For the fleshy part being thick and powerful will prevail over the bandaging, and not be overcome by it. To come to our subject, one should extend very strongly and without deviation leaving no deficiency, for the disgrace and harm are great if the result is a shortened thigh. The arm, indeed, when shortened may be concealed and the fault is not great, but the leg when shortened will leave the patient lame, and the sound leg being longer [by comparison] exposes the defect; so that if a patient is going to have unskilful treatment, it is better that both his legs should be broken than one of them, for then at least he will be in equilibrium. When, therefore, you have made sufficient extension, you should adjust the parts with the palms of the hands and bandage in the same way as was described before, placing the head of the bandage as

directed and carrying it upwards. And he should give the same answers as before, and experience the same trouble and relief. Let the change of dressing be made in the same way, and the same application of splints. The thigh-bone gets firm in forty days.

20. One should also bear the following in mind, that the thigh-bone is curved outwards rather than inwards, and to the front rather than to the back, so it gets distorted in these directions if not skilfully treated. Furthermore it is less covered with flesh on these parts so that distortions cannot be hidden. If, then, you suspect anything of this kind, you should have recourse to the mechanical methods recommended for distortion of the upper arm. Some additional turns of bandage should be made round the hip and loins so that the groins and the joint at the so-called fork may be included, for besides other benefits, it prevents the ends of the splints from doing damage by contact with the uncovered parts. The splints should always come considerably short of the bare part at either end, and care should always be taken as to their position so that it is neither on the bone where there are natural projections about the joint, nor on the tendon.

21. As to the swellings which arise owing to pressure behind the knee or at the foot or elsewhere, dress them with plenty of crude wool, well pulled out, sprinkling it with oil and wine, after anointing with cerate, and if the splints cause pressure relax them at once. You will reduce the swellings by applying slender bandages after removing the splints, beginning from the lowest part and passing upwards, for so the swelling would be most rapidly reduced and flow back above the original dressing. But you should not use this method of bandaging unless there is danger of blisters forming or mortification at the swelling. Now, nothing of this kind happens unless one puts great pressure on the fracture, or the part is kept hanging down or is scratched with the hand, or some other irritant affects the skin.

22. As to a hollow splint, if one should pass it under the thigh itself and it does not go below the bend of the knee it would do more harm than good; for it would prevent neither

the body nor the leg from moving apart from the thigh, would cause discomfort by pressing against the flexure of the knee, and incite the patient to bend the knee, which is the last thing he should do. For when the thigh and leg are bandaged, he who bends the knee causes all sorts of disturbance to the dressings, since the muscles will necessarily change their relative positions and there will also necessarily be movement of the fractured bones. Special care, then, should be taken to keep the knee extended. I should think that a hollow splint reaching [evenly?] from hip to foot would be useful, especially with a band passed loosely round at the knee to include the splint, as babies are swaddled in their cots. Then if the thigh-bone is distorted upwards [i.e. forwards] or sideways it will thus be more easily controlled by the hollow splint. You should, then, use the hollow splint for the whole limb or not at all.

23. In fractures both of the leg and of the thigh great care should be taken that the point of the heel is in good position. For if the foot is in the air while the leg is supported, the bones at the shin necessarily present a convexity, while if the foot is propped up higher than it should be, and the leg imperfectly supported, this bone in the shin part has a more hollow appearance that the normal, especially if the heel happens to be large compared with the average in man. So, too, all bones solidify more slowly if not placed in their natural position and kept at rest in the same posture, and the callus is weaker.

24. The above remarks apply to those whose bones are fractured without protrusion or wound of other kind. In fractures with protrusion, where they are single and not splintered, if reduced on the same or following day, the bones keeping in place, and if there is no reason to expect elimination of splinters, or even cases in which, though there is an external wound, the broken bones do not stick out, nor is the nature of the fracture such that any splinters are likely to come to the surface: in such cases they do neither much good nor much harm who treat the wound with a cleansing plaster, either pitch cerate, or an application for fresh wounds, or whatever

else they commonly use, and bind over it compresses soaked in wine, or uncleansed wool or something of the kind. And after the wounds are cleansed and already united, they attempt to make adjustment with splints and use a number of bandages. This treatment does some good and no great harm. The bones, however, cannot be so well settled in their proper place, but become somewhat unduly swollen at the point of fracture. If both bones are broken, either of forearm or leg, there will also be shortening.

25. Then there are others who treat such cases at once with bandages, applying them on either side, while they leave a vacancy at the wound itself and let it be exposed. Afterwards, they put one of the cleansing applications on the wound, and treat it with pads steeped in wine, or with crude wool. This treatment is bad, and those who use it probably show the greatest folly in their treatment of other fractures as well as these. For the most important thing is to know the proper way of applying the head of the bandage, and how the chief pressure should be made, also what are the benefits of proper application and of getting the chief pressure in the proper place, and what is the harm of not placing the bandage rightly, and of not making pressure where it should chiefly be, but at one side or the other. Now the results of each were explained in what has been written above. The treatment, too, is itself evidence; for in a patient so bandaged the swelling necessarily arises in the wound itself, since if even healthy tissue were bandaged on this side and that, and a vacancy left in the middle, it would be especially at the vacant part that swelling and decoloration would occur. How then could a wound fail to be affected in this way? For it necessarily follows that the wound is discoloured with everted edges, and has a watery discharge devoid of pus,* and as to the bones, even those which were not going to come away do come away. The wound will become heated and throbbing, and they are obliged to put on an additional plaster because of the swelling; and this too will be harmful to patients bandaged at either side of the wound, for an unprofitable burden is added to the throbbing.

* That is, an unhealthy discharge without 'purification'.

They finally take off the dressings, when they find there is aggravation, and treat it for the future without bandaging. Yet none the less, if they get another wound of the same sort, they use the same treatment, for they do not suppose that the outside bandaging and exposure* of the wound is to blame, but some mishap. However, I should not have written so much about this had I not known well the harmfulness of this dressing and that many use it; and that it is of vital importance to unlearn the habit. Besides, it is an evidence of the truth of what was written before on the question whether the greatest or least pressure should come at the fracture.†

26. [Proper treatment of compound fractures.] To speak summarily, when there is no likelihood of elimination of bone, one should use the same treatment as in cases of fracture without external wound. The extensions and adjustments of the bones should be made in the same way, and so too with the bandaging. After anointing the wound itself with pitch cerate, bind a thin doubled compress over it, and anoint the surrounding parts with a thin layer of cerate. The bandages and other dressings should be torn in rather broader strips than if there was no wound, and the one first used should be a good deal wider than the wound; for bandages narrower than the wound bind it like a girdle, which should be avoided; rather let the first turn take in the whole wound, and let the bandage extend beyond it on both sides. One should, then, put the bandage just in the line of the wound, make rather less pressure than in cases without a wound, and distribute the dressing as directed above. The bandages should always be of the pliant kind, and more so in these cases than if there was no wound. As to number, let it not be less than those mentioned before, but even a little greater. When the bandaging is finished it should appear to the patient to be firm without pressure, and

*Exposure here cannot mean exposure to cold or even bareness – the foolish surgeons cover the wound with wool or pads – it means absence of due pressure, the proper graduation of which is the main point in Hippocratic bandaging.

†According to Adams, who translated Hippocrates in 1849, this warning was still necessary in his time.

he should say that the greatest firmness is over the wound. There should be the same periods of a sensation of greater firmness, and greater relaxation, as were described in the former cases. Change the dressings every other day, making the changes in similar fashion except that, on the whole, the pressure should be less in these cases. If the case takes a natural course according to rule, the part about the wound will be found progressively diminished and all the rest of the limb included in the bandage will be slender. Purification* will take place more rapidly than in wounds treated otherwise, and all fragments of blackened or dead tissue are more rapidly separated and fall off under this treatment than with other methods. The wound, too, advances more quickly to cicatrization thus than when treated otherwise. The cause of all this is that the wound and the surrounding parts become free from swelling. In all other respects, then, one should treat these cases like fractures without a wound, but splints should not be used.† This is why the bandages should be more numerous than in the other cases both because there is less pressure and because the splints are applied later. But if you do apply splints, do not put them in the line of the wound; especially apply them loosely, taking care that there is no great compression from the splints. This direction was also given above. Diet, however, should be more strict and kept up longer in cases where there is a wound from the first and where the bones protrude, and on the whole, the greater the injury the more strict and prolonged should be the dieting.

27. The same treatment of the wounds applies also to cases of fracture which are at first without wound, but where one occurs during treatment either through too great compression by bandages or the pressure of a splint or some other cause. In such cases the occurrence of ulceration is recognized by pain and throbbing: also the swelling on the extremities gets harder, and if you apply the finger the redness is removed but quickly returns. So, if you suspect anything of this kind you should undo the dressings, if there is irritation below the

---

*i.e. discharge of laudable pus.
†We must evidently understand 'so soon'.

under bandages, or in the rest of the bandaged part, and use pitch cerate instead of the other plaster. Should there be none of this, but the sore itself is found to be irritated, extensively blackened or foul with tissues about to suppurate and tendons on the way to be thrown off, it is by no means necessary to leave them exposed, or to be in any way alarmed at these suppurations, but treat them for the future in the same manner as cases in which there is a wound from the first. The bandaging should begin from the swelling at the extremities and be quite slack; then it should be carried right on upwards, avoiding pressure in any place, but giving special support at the wound and decreasing it elsewhere. The first bandages must be clean and not narrow, their number as many as when splints are applied or a little fewer. On the wound itself a compress anointed with white cerate is sufficient; for if flesh or tendon be blackened it will also come away. One should treat such cases not with irritant, but with mild applications, just like burns. Change the dressing every other day but do not apply splints. Keep the patient at rest and on low diet even more than in the former case. One should know if either flesh or tendon is going to come away that the loss will be much less extensive and will be brought about much quicker, and the surrounding parts will be much less swollen [by this treatment], than if on removing the bandage one applied some detersive plaster to the wound. Besides, when the part that is going to suppurate off does come away, flesh formation and cicatrization will be more rapid with the former treatment than with any other. The whole point is to know the correct method and due measure in dressing these cases. Correctness of position also contributes to the result, as well as diet and the suitability of the bandages.

28. If, perchance, you are deceived in fresh cases, and think there will be no elimination of bones, yet they show signs of coming to the surface, the use of the above mode of treatment need not cause alarm, for no great damage will be done if only you have sufficient manual skill to apply the dressings well and in a way that will do no harm. The following is a sign of approaching elimination of bone in a case thus treated. A

large amount of pus flows from the wound, which appears turgid. So the dressing should be changed more often because of the soaking,* for thus especially they get free from fever, if there is no great compression by the bandages, and the wound and surrounding parts are not engorged. But separations of very small fragments require no great alteration of treatment beyond either loose bandaging so as not to intercept the pus but allow it to flow away freely; or even more frequent change of dressing till the bone separates, and no application of splints.

29. But in cases where separation of a rather large bone is probable, whether you prognosticate it from the first or recognize it later, the treatment should not be the same, but, while the extensions and adjustments should be done as was directed, the compresses should be double, half a span in breadth at least – take the nature of the wound as standard for this – and in length a little less than will go twice round the wounded part, but a good deal more than will go once round. Provide as many of these as may suffice, and after soaking them in dark astringent wine, apply them beginning from their middle as is done with a two-headed under-bandage; enveloping the part and then leaving the ends crossed oblique-ly, as with the adze-shaped bandage. Put them both over the wound itself and on either side of it, and though there should be no compression, they should be applied firmly so as to support the wound. On the wound itself one should put pitch cerate or one of the applications for fresh injuries or any other appropriate remedy which will serve as an embrocation. If it is summer time soak the compresses frequently with wine, but if winter apply plenty of crude wool moistened with wine and oil. A goat's skin should be spread underneath to make free course for discharges, giving heed to drainage and bearing in mind that these regions (when patients lie a long time in the same posture) develop sores difficult to heal.

30. As to cases which cannot be treated by bandaging in one of the ways which have been or will be described, all the more care should be taken that they shall have the fractured

*'Maceration', 'abundance of humours'.

limb in good position in accord with its normal lines, seeing to it that the slope is upwards rather than downwards. If one intends to do the work well and skilfully, it is worth while to have recourse to mechanism, that the fractured part may have proper but not violent extension. It is especially convenient to use mechanical treatment for the leg. Now, there are some who in all cases of leg fractures, whether they are bandaged or not, fasten the foot to the bed, or to some post which they fix in the ground by the bed. They do all sorts of harm and no good; for extension is not ensured by fastening the foot, since the rest of the body will none the less move towards the foot, and thus extension will not be kept up. Nor is it of any use for preserving the normal line, but even harmful. For when the rest of the body is turned this way or that, the ligature in no way prevents the foot and the bones connected with it from following the movement: if it were not tied up, there would be less distortion, for it would not be left behind so much in the movement of the rest of the body. Instead of this, one should get two rounded circlets sewn in Egyptian leather such as are worn by those who are kept a long time shackled in the large fetters. The circlets should have coverings on both sides deeper on the side facing the injury and shallower on that facing the joints. They should be large and soft, fitting the one above the ankle, the other below the knee. They should have on each side two attachments of leather thongs, single or double, short like loops, one set at the ankle on either side, the other on either side of the knee (and the upper circlet should have others like them in the same straight line, i.e. just opposite those below). Then take four rods of cornel wood of equal size, the thickness of a finger; and of such length as when bent they fit into the appendices, taking care that the ends of the rods do not press upon the skin but on the projecting edges of the circlet. There should be three or more pairs of rods, some longer than the others and some shorter and more slender, so as to exert greater or less tension at pleasure. Let the rods be placed separately on either side of the ankles. This mechanism if well arranged will make the extension both correct and even in accordance with the normal lines, and cause no pain in the wound, for the outward pressure, if there is any, will be diverted partly to the foot and partly to the thigh. The rods are better placed, some on one side and some on the other side of the ankles, so as not to interfere with the position of the leg; and the wound is both easy to examine and easy to handle. For, if one pleases, there is nothing to prevent the two upper rods from being tied together, so that, if one wants to put something lightly over it, the covering is kept up away from the wound. If then the circlets are supple, of good quality, soft and newly sewn, and the extension by the bent rods suitably regulated as just described, the mechanism is of good use, but if any of these things are not well arranged it will harm rather than help. Other mechanisms also should either be well arranged or not used, for it is shameful and contrary to the art to make a machine and get no mechanical effect.

31. Again, most practitioners treat fractures, whether with or without wounds, by applying uncleansed wool during the first days, and this appears in no way contrary to the art. Those who because they have no bandages are obliged to get wool for first-aid treatment are altogether excusable, for in the absence of bandages one would have nothing much better than wool with which to dress such cases; but it should be plentiful, well pulled out and not lumpy; if small in amount and of poor quality its value is also small. Now, those who think it correct to dress with wool for one or two days, and on the third or fourth day use bandages with compression and extension just at this period are very ignorant of the healing art, and that on a most vital point. For, to speak summarily, the third or fourth day is the very last on which any lesion should be actively interfered with; and all probings as well as everything else by which wounds are irritated should be avoided on these days. For, as a rule, the third or fourth day sees the birth of exacerbations in the majority of lesions, both where the tendency is to inflammation and foulness, and in those which turn to fever. And if any instruction is of value this is very much so. For what is there of most vital importance in the healing art to which it does not apply, not only as regards wounds but

many other maladies? Unless one calls all maladies wounds, for this doctrine also has reasonableness, since they have affinity one to another in many ways. But those who think it correct to use wool till seven days are completed and then proceed to extension, coaptation and bandaging would appear not so unintelligent, for the most dangerous time for inflammation is past, and the bones after this period will be found loose and easy to put in place. Still, even this treatment is much inferior to the use of bandages from the beginning, for that method results in the patients being without inflammation on the seventh day and ready for complete dressing with splints, while the former one is much slower, and has some other disadvantages; but it would take long to describe everything.

In cases where the fractured and projecting bones cannot be settled into their proper place, the following is the method of reduction. One must have iron rods made in fashion like the levers used by stone masons, broader at one end and narrower at the other. There should be three and even more that one may use those most suitable. Then one should use these, while extension is going on, to make leverage, pressing the under side of the iron on the lower bone, and the upper side against the upper bone, in a word just as if one would lever up violently a stone or log. The irons should be as strong as possible so as not to bend. This is a great help, if the irons are suitable and the leverage used properly; for of all the apparatus contrived by men these three are the most powerful in action – the wheel and axle, the lever and the wedge. Without some one, indeed, or all of these, men accomplish no work requiring great force. This lever method, then, is not to be despised, for the bones will be reduced thus or not at all. If, perchance, the upper bone over-riding the other affords no suitable hold for the lever, but being pointed, slips past, one should cut a notch in the bone to form a secure lodgement for the lever. The leverage and extension should be done on the first or second day, but not on the third, and least of all on the fourth and fifth. For to cause disturbance without reduction on these days would set up inflammation, and no less so if there was reduction; spasm, indeed, would much more likely be caused if reduction succeeded than if it failed. It is well to know this, for if spasm supervenes after reduction there is not much hope of recovery. It is advantageous to reproduce the displacement, if it can be done without disturbance, for it is not when parts are more relaxed than usual that spasms and tetanus supervene, but when they are more on the stretch. As regards our subject, then, one should not disturb the parts on the days above mentioned, but study how best to oppose inflammation in the wound and favour suppuration. At the end of seven days, or rather more, if the patient is free from fever and the wound not inflamed, there is less objection to an attempt at reduction, if you expect to succeed; otherwise you should not give the patient and yourself useless trouble.

32. The proper modes of treatment after you reduce the bones to their place have already been described, both when you expect bones to come away and when you do not. Even when you expect bones to come away you should use in all such cases the method of separate bandages, as I said, beginning generally with the middle of the bandage as when an under-bandage is applied from two heads. Regulate the process with a view to the shape of the wound that it may be as little as possible drawn aside or everted by the bandaging: for in some cases it is appropriate to bandage to the right, in others to the left, in others from two heads.

33. As to bones which cannot be reduced, it should be known that just these will come away, as also will those which are completely denuded. In some cases the upper part of the bones are denuded, in others the soft parts surrounding them perish, and the starting point of the necrosis is, in some of the bones, the old wound, in others not. It is more extensive in some and less so in others, and some bones are small, others large. It follows from the above that one cannot make a single statement as to when the bones will come away, for some separate sooner owing to their small size, others because they come at the end [of the fracture] while others do not come away [as wholes] but are exfoliated after desiccation and corruption. Besides this, the treatment makes a difference. As a general rule, bones are most quickly eliminated in cases where

suppuration is quickest, and the growth of new flesh most rapid and good; for it is the growth of new flesh in the lesion that as a rule lifts up the fragments. As to a whole circle of bone, if it comes away in forty days it will be a good separation, for some cases go on to sixty days or even more. The more porous bones come away more quickly, the more solid more slowly; for the rest, the smaller ones take much less time, and so variously. The following are the indications for resection of a protruding bone: if it cannot be reduced, but only some small portion seems to come in the way, and it is possible to remove it; if it is harmful, crushing some of the tissues, and causing wrong position of the part, and if it is denuded, this also should be removed. In other cases it makes no great difference whether there is resection or not. For one should bear clearly in mind that when bones are entirely deprived of soft parts and dried up they will all come away completely: and one should not resect those bones which are going to be exfoliated. Draw your conclusion as to bones which will come away completely from the symptoms set forth.

34. Treat such cases with compresses and vinous applications as described above in the case of bones about to be eliminated. Take care not to moisten with cold fluids at first, for there is risk of feverish rigors and further risk of spasms, for cold substances provoke spasms and sometimes ulcerations. Bear in mind that there must be shortening of the parts in cases where, when both bones are broken, they are treated while overlapping, also in cases where the circle of bone is eliminated entire.

35. Cases where the bone of the thigh or upper arm protrudes rarely recover; for the bones are large and contain much marrow, while the cords, muscles and blood-vessels which share in the injury are numerous and important. Besides, if you reduce the fracture, convulsions are liable to supervene, while in cases not reduced there are acute bilious fevers with hiccough and mortification. Cases where reduction has not been made or even attempted are no less likely to recover, and recovery is more frequent when the lower than when

the upper part of the bone projects. There may be survival even in cases where reduction is made, but it is rare indeed. There are great differences between one way of dealing with the case and another, and between one bodily constitution and another as to power of endurance. It also makes a great difference whether the bone protrudes on the inner or outer side of the arm or thigh, for many important blood-vessels stretch along the inner side, and lesions of some of them are fatal; there are also some on the outside, but fewer. In such injuries, then, one must not overlook the dangers or the nature of some of them, but foretell them as suits the occasion. If you have to attempt reduction and expect to succeed and there is no great overriding of the bone, and the muscles are not retracted (for they are wont to retract) leverage combined with extension would be well employed even in these cases.

36. After reduction one should give a mild dose of hellebore on the first day, if it is reduced on the first day, otherwise one should not even attempt it. The wound should be treated with the remedies used for the bones of a broken head. Apply nothing cold and prescribe entire abstinence from solid food. If he is of a bilious nature give him a little aromatic hydromel sprinkled in water, but if not, use water as beverage. And if he is continuously feverish keep him on this regimen for fourteen days at least, but if there is no fever, for seven days, then return by a regular gradation to ordinary diet. In cases where the bones are not reduced, a similar purgation should be made and so with the management of the wounds and the regimen. Likewise do not stretch the unreduced part, but even bring it more together so that the seat of the wound may be more relaxed. Elimination of the bones takes time, as was said before. One should especially avoid such cases if one has a respectable excuse, for the favourable chances are few, and the risks many. Besides, if a man does not reduce the fracture, he will be thought unskilful, while if he does reduce it he will bring the patient nearer to death than to recovery.

37. Dislocations at the knee and disturbances of the bones are much milder than displacements and dislocations at the elbow; for the articular end of the thigh-bone is more compact

in relation to its size than is that of the arm-bone, and it
alone has a regular conformation, a rounded one, whereas the
articular end of the humerus is extensive, having several cavities. Besides this the leg-bones are about the same size, the
outer one overtops the other to some little extent not worth
mention,* and opposes no hindrance to any large movement
though the external tendon of the ham arises from it. But
the bones of the forearm are unequal, and the shorter [radius]
much the thicker, while the more slender one [ulna] goes far
beyond and overtops the joint. This, however, is attached to
the ligaments at the common junction of the bones.† The
slender bone has a larger share than the thicker one of the
attachments of ligaments in the arm. Such then is the disposition of these articulations and of the bones of the elbow.
Owing to the way they are disposed the bones at the knee are
often dislocated‡ but easily put in, and no great inflammation
or fixation of the joint supervenes. Most dislocations are inwards,§ but some outwards and some into the knee flexure.
Reduction is not difficult in any of these cases: as to external
and internal dislocations, the patient should be seated on the
ground or something low, and have the leg raised, though
not greatly. Moderate extension as a rule suffices; make extension on the leg and counter-extension on the thigh.

38. Dislocations at the elbow are more troublesome than
those at the knee, and harder to put in, both because of the
inflammation and because of the conformation of the bones,
unless one puts them in at once. It is true that they are more
rarely dislocated than the above, but they are harder to put up,
and inflammation and excessive formation of callus is more
apt to supervene.

39. [Dislocation of radius.] The majority of these are small
displacements sometimes inwards, towards the side and ribs,

*A curious error, perhaps due to an effort to make the fibula resemble
the ulna as far as possible. (The fibula does not reach the top of the tibia.)

† The ulna is attached to the ligaments of the elbow joint, at the point
where it joins the radius.

‡A strange remark, perhaps includes displacement of the kneecap.
Displacements of cartilages are not noticed.

§ Of the thigh-bone.

sometimes outwards [our 'forwards' and 'backwards']. The joint is not dislocated as a whole, but maintaining the connection with the cavity of the humerus, where the projecting part of the ulna sticks out. Such cases, then, whether dislocation is to one side or the other, are easy to reduce, and direct extension in the line of the upper arm is quite enough; one person may make traction on the wrist, another does so by clasping the arm at the axilla, while a third presses with the palm of one hand on the projecting part and with the other makes counter-pressure near the joint.

40. Such dislocations yield readily to reduction if one reduces them before they are inflamed; the dislocation is usually rather inwards [forwards], but may also be outwards, and is easily recognized by the shape. And they are often reduced even without vigorous extension. In the case of internal dislocations one should push the joint back into its natural place, and turn the forearm rather towards the prone position. Most dislocations of the elbow are of this kind.

41. [Complete dislocation of the elbow backwards and forwards.] If the articular end of the humerus passes either this way or that over the part of the ulna which projects into its cavity (the latter* indeed occurs rarely, if it does occur), extension in the line of the limb is no longer equally suitable, for the projecting part of the ulna prevents the passage of the humerus. In patients with these dislocations, extension should be made after the manner which has been described above for putting up a fractured humerus. Make traction upwards from the armpit, and apply pressure downwards at the elbow itself, for this is the most likely way to get the humerus lifted above its own socket, and if it is so raised, replacement by the palms of hands is easy, using pressure with one hand to put in the projecting part of the humerus, and making counter-pressure on the ulna at the joint to put it back. The same method suits both cases. This has, indeed, less claim to be called the most regular method of extension in such a dislocation and reduction would also be made by direct extension, but less easily.

42. [Internal lateral distortion of the forearm.] Suppose the

*Refers to 'backwards', which can hardly occur without fracture.

humerus to be dislocated forwards. This happens very rarely; but what might not be dislocated by a sudden violent jerk? For many other bones are displaced from their natural position, though the opposing obstacle may be great. Now, there is a great obstacle to this jerking out, namely the passage over the thicker bone [radius] and the extensive stretching of the ligaments, but nevertheless it is jerked out in some cases. Symptoms in cases of such jerkings out: they cannot bend the elbow at all, and palpation of the joint makes it clear. If, then, it is not reduced at once, violent and grave inflammation occurs with fever, but if one happens to be on the spot it is easily put in. One should take a hard bandage (a hard rolled bandage of no great size is sufficient) and put it crosswise in the bend of the elbow, suddenly flex the elbow, and bring the hand as close as possible to the shoulder. This mode of reduction is sufficient for such jerkings out. Direct extension, too, can accomplish this reduction. One must, however, use the palms, putting one on the projecting part of the humerus at the elbow and pushing backwards [our 'inwards'], and with the other making counter-pressure below the point of the elbow, inclining the parts into the line of the ulna. In this form of dislocation, the mode of extension described above as proper to be used in stretching the fractured humerus when it is going to be bandaged is also effective. And when extension is made, application of the palms should be made as described above.

43. [External lateral dislocation of forearm.] If the humerus is dislocated backwards [our 'inwards'] – this occurs rarely, and is the most painful of all, most frequently causing continuous fever with vomiting of pure bile, and fatal in a few days – the patient cannot extend the arm. If you happen to be quickly on the spot, you ought to extend the elbow forcibly, and it goes in of its own accord. But if he is feverish when you arrive, do not reduce, for the pain of a violent operation would kill him. It is a general rule not to reduce any joint when the patient has fever, least of all the elbow.

44. [Separation of radius.] There are also other troublesome lesions of the elbow. Thus the thicker bone is sometimes separated from the other, and they can neither flex nor extend

the joint as before. The lesion is made clear by palpation at the bend of the elbow about the bifurcation of the blood-vessel* which passes upwards along the muscle.† In such cases it is not easy to bring the bone into its natural place, for no symphysis of two bones when displaced is permanently settled in its old position, but the diastasis [separation] necessarily remains as a swelling. How a joint ought to be bandaged was described in the case of the ankle.

45. [Fractures of olecranon.] There are cases in which the bone of the forearm [ulna] is fractured where it is subjacent to the humerus, sometimes the cartilaginous part from which the tendon at the back of the arm arises, sometimes the part in front at the origin of the anterior coronoid process, and when this occurs it is complicated with fever and dangerous, though the joint [articular end of humerus] remains in its place, for its entire base comes above this bone.‡ But when the fracture is in the place on which the articular head of the humerus rests, the joint becomes more mobile if it is a complete cabbage-stalk fracture [i.e. right across]. Speaking generally, fractures are always less troublesome than cases where no bones are broken, but there is extensive contusion of blood-vessels and important cords in these parts. For the latter lesions involve greater risk of death than do the former, if one is seized with continued fever. Still, fractures of this kind rarely occur.

46. Sometimes the actual head of the humerus is fractured at the epiphysis, but this, though apparently a very grave lesion, is much milder than injuries of the elbow joint.

47. How, then, each dislocation is most appropriately [reduced and] treated has been described; especially the value of immediate reduction owing to the rapid inflammation of the ligaments. For, even when parts that are put out are put in at once, the tendons are apt to become contracted and to hinder for a considerable time the natural amount of flexion and extension. All such lesions, whether avulsions, separations or

---

*Cephalic vein.   †Biceps.

‡ The articular end of the humerus rests entirely on the olecranon, the arm being bent.

dislocations, require similar treatment, for they should all be treated with a quantity of bandages, compresses and cerate, as with fractures. The position of the elbow should in these cases, too, be the same in all respects as in the bandaging of patients with fractured arm or forearm; for this position is most generally used for all the dislocations, displacements and fractures, and is also most useful as regards the future condition, in respect both of extension and flexion in the several cases, since from it the way is equally open in both directions. This attitude is also most easily kept up or returned to by the patient himself. And besides this, if ankylosis should prevail, an arm ankylosed in the extended position would be better away, for it would be a great hindrance and little use. If flexed, on the other hand, it would be more useful, and still more useful if the ankylosis occurred in an attitude of semi-flexion. So much concerning the attitude.

48. One should bandage by applying the head of the first roll to the place injured whether it be fractured, dislocated, or separated. The first turns should be made there and the firmest pressure, slackening off towards each side. The bandaging should include both fore and upper arm, and be carried much further each way than most practitioners do, that the oedema may be repelled as far as possible from the lesion to either side. Let the point of the elbow be also included in the bandage, whether the lesion be there or not, that the oedema may not be collected about this part. One should take special care in the dressing that, so far as possible, there shall be no great accumulation of bandage in the bend of the elbow, and that the firmest pressure be made at the lesion. For the rest, let him deal with the case as regards pressure and relaxation, in the same way, and according to the same respective periods, as was previously described in the treatment of fractured bones. Let the change of dressings take place every third day, and he should feel them relaxed on the third day, as in the former case. Apply the splints at the proper time – for their use is not unsuitable whether there is fracture or not, if there is no fever – but they should be applied as loosely as possible, those of the

arm being under and those of the forearm on the top.* The splints should not be thick, and must be unequal in length in order to overlap one another where it is convenient, judging by the degree of flexion. So, too, as regards the application of compresses, one should follow the directions for the splints. They should be rather thicker at the point of lesion. The periods are to be estimated by the inflammation and the directions already given.

*Hippocrates had no angular splints, and straight ones applied to the bent arm above and below the elbow had to be so arranged that one set overlapped the other at the sides.

EMBRYOLOGY AND ANATOMY

TRANSLATED BY I. M. LONIE

THE SEED and THE NATURE
OF THE CHILD*

These two treatises, together with the work known as Diseases IV, *form the group of so-called embryological treatises. Cross-references within them imply that they were written by the same author, and many scholars have supposed that they were composed as a unity. Their chief interest lies in the insight they provide into one fourth-century writer's views on the growth of the embryo and on the problems of generation, heredity and sex differentiation.*

THE SEED

1. *All things are governed by law.* The sperm of the human male comes from all the fluid in the body: it consists of the most potent part of this fluid, which is secreted from the rest. The evidence that it is the most potent part which is secreted is the fact that even though the actual amount we emit in intercourse is very small, we are weakened by its loss. What happens is this: there are veins and nerves which extend from every part of the body to the penis. When as the result of gentle friction these vessels grow warm and become congested, they experience a kind of irritation, and in consequence a feeling of pleasure and warmth arises over the whole body. Friction on the penis and the movement of the whole man cause the fluid in the body to grow warm: becoming diffuse and agitated by the movement it produces a foam, in the same way as all other fluids produce foam when they are agitated. But in the case of the human being what is secreted as foam is the most potent and the richest part of the fluid. This fluid is diffused from the brain into the loins and the whole body, but in particular into the spinal marrow: for passages extend into this from the whole body, which enable the fluid to pass to and

*The text translated is one which is to appear in a new edition with translation and commentary being prepared by I. M. Lonie and Gerhard Baader for the series *Ars Medica*, published by Walter De Gruyter and Co., Berlin.

from the spinal marrow. Once the sperm has entered the spinal marrow it passes in its course through the veins along the kidneys (sometimes if there is a lesion of the kidneys, blood is carried along with the sperm). From the kidneys it passes via the testicles into the penis – not however by the urinary tract, since it has a passage of its own which is next to the urinary tract.

Those who have nocturnal emissions have them for the following reasons: when the fluid in the body becomes diffuse and warmed throughout – whether through fatigue or through some other cause – it produces foam. As this is secreted, the man sees visions as though he were having intercourse, for the fluid is precisely the same as that which is emitted in intercourse. However, erotic dreams and the nature and effects of this whole complaint, and why it is a precursor of insanity, are no part of my present subject. So much then for that.

2. The reason that eunuchs do not have intercourse is that their spermal passage is destroyed. The passage lies through the testicles themselves. Moreover, the testicles are connected to the penis by a mass of slender ligaments, which raise and lower the penis. These are cut off in the operation, and that is why eunuchs are impotent. In the case of those whose testicles are crushed, the spermal passage is blocked, for the testicles are damaged; and the ligaments, becoming calloused and insensitive as a result of the damage, are no longer able to tighten and relax. Those on the other hand who have had an incision made by the ear, can indeed have intercourse and emit sperm, but the amount is small, weak and sterile. For the greater part of the sperm travels from the head past the ears into the spinal marrow: now when the incision has formed a scar, this passage becomes obstructed. In the case of children their vessels are narrow and filled, and therefore prevent the passage of sperm, so that the irritation cannot occur as it does in the adult. Hence the fluid in the body cannot be agitated sufficiently to secrete sperm. Girls while they are still young do not menstruate for the same reason. But as both boys and girls grow, the vessels which extend in the boy's case to the penis and in the girl's to the womb become permeable and open up in the

process of growth; a way is opened up through the narrow passages, and the fluid, finding sufficient space, can become agitated. That is why when they reach puberty, sperm can flow in the boy and the menses in the girl. Such is my explanation of these facts.

3. The sperm is, as I say, secreted from the whole body – from the hard parts as well as from the soft, and from the total bodily fluid. This fluid has four forms: blood, bile, water and phlegm. All four are innate in man, and they are the origin of disease. (I have already discussed these forms, and why both diseases and their resolutions come from them.)* I have now dealt with the subject of sperm: its origin, how and why it originates, and in the case of those who do not have sperm, why this is so; and I have dealt with menstruation in girls.

4. In the case of women, it is my contention that when during intercourse the vagina is rubbed and the womb is disturbed, an irritation is set up in the womb which produces pleasure and heat in the rest of the body. A woman also releases something from her body, sometimes into the womb, which then becomes moist, and sometimes externally as well, if the womb is open wider than normal. Once intercourse has begun, she experiences pleasure throughout the whole time, until the man ejaculates. If her desire for intercourse is excited, she emits before the man, and for the remainder of the time she does not feel pleasure to the same extent; but if she is not in a state of excitement, then her pleasure terminates along with that of the man. What happens is like this: if into boiling water you pour another quantity of water which is cold, the water stops boiling. In the same way, the man's sperm arriving in the womb extinguishes both the heat and the pleasure of the woman. Both the pleasure and the heat reach their peak simultaneously with the arrival of the sperm in the womb, and then they cease. If, for example, you pour wine on a flame, first of all the flame flares up and increases for a short period when you pour the wine on, then it dies away. In the same way the woman's heat flares up in response to the man's sperm, and

* The author refers to his treatise, now called *Diseases* IV, which Littré prints as a continuation of *The Nature of the Child*.

then dies away. The pleasure experienced by the woman during intercourse is considerably less than the man's, although it lasts longer. The reason that the man feels more pleasure is that the secretion from the bodily fluid in his case occurs suddenly, and as the result of a more violent disturbance than in the woman's case.

Another point about women: if they have intercourse with men their health is better than if they do not. For in the first place, the womb is moistened by intercourse, whereas when the womb is drier than it should be it becomes extremely contracted, and this extreme contraction causes pain to the body. In the second place, intercourse by heating the blood and rendering it more fluid gives an easier passage to the menses; whereas if the menses do not flow, women's bodies become prone to sickness. I shall explain why this is so in my course on women's diseases.* So much then for that subject.

5. When a woman has intercourse, if she is not going to conceive, then it is her practice to expel the sperm produced by both partners whenever she wishes to do so. If however she is going to conceive, the sperm is not expelled, but remains in the womb. For when the womb has received the sperm it closes up and retains it, because the moisture causes the womb's orifice to contract. Then both what is provided by the man and what is provided by the woman is mixed together. If the woman is experienced in matters of childbirth, and takes note when the sperm is retained, she will know the precise day on which she has conceived.

6. Now here is a further point. What the woman emits is sometimes stronger, and sometimes weaker; and this applies also to what the man emits. In fact both partners alike contain both male and female sperm (the male being stronger than the female must of course originate from a stronger sperm). Here is a further point: if (a) both partners produce a stronger sperm, then a male is the result, whereas if (b) they produce a weak form, then a female is the result. But if (c) one partner produces one kind of sperm, and the other another, then the

*Cf. ch. 15. The work referred to corresponds in part to the treatise *The Diseases of Women.*

resultant sex is determined by whichever sperm prevails in quantity. For suppose that the weak sperm is much greater in quantity than the stronger sperm: then the stronger sperm is overwhelmed and, being mixed with the weak, results in a female. If on the contrary the strong sperm is greater in quantity than the weak, and the weak is overwhelmed, it results in a male. It is just as though one were to mix together beeswax with suet, using a larger quantity of the suet than of the beeswax, and melt them together over a fire. While the mixture is still fluid, the prevailing character of the mixture is not apparent: only after it solidifies can it be seen that the suet prevails quantitatively over the wax. And it is just the same with the male and female forms of sperm.

7. Now that both male and female sperm exist in both partners is an inference which can be drawn from observation. Many women have borne daughters to their husbands and then, going with other men, have produced sons. And the original husbands – those, that is, to whom their wives bore daughters – have as the result of intercourse with other women produced male offspring; whereas the second group of men, who produced male offspring, have with yet other women produced female offspring. Now this consideration shows that both the man and the woman have male and female sperm. For in the partnership in which the women produced daughters, the stronger sperm was overwhelmed by the larger quantity of the weaker sperm, and females were produced; while in the partnership in which these same women produced sons, it was the weak which was overwhelmed, and males were produced. Hence the same man does not invariably emit the strong variety of sperm, nor the weak invariably, but sometimes the one and sometimes the other; the same is true in the woman's case. There is therefore nothing anomalous about the fact that the same women and the same men produce both male and female sperm: indeed, these facts about male and female sperm are also true in the case of animals.

8. Sperm is a product which comes from the whole body of each parent, weak sperm coming from the weak parts, and strong sperm from the strong parts. The child must necessarily

correspond. If from any part of the father's body a greater quantity of sperm is derived than from the corresponding part of the mother's body the child will, in that part, bear a closer resemblance to its father; and vice versa. The following cases however are impossible: (a) the child resembles its mother in all respects, and its father in none; (b) the child resembles its father in all respects, and its mother in none; (c) the child resembles neither parent in any respect. No: it must inevitably resemble each parent in some respect, since it is from both parents that the sperm comes to form the child. The child will resemble in the majority of its characteristics that parent who has contributed a greater quantity of sperm to the resemblance, and from a greater number of bodily parts. And so it sometimes happens that although the child is a girl she will bear a closer resemblance in the majority of her characteristics to her father than to her mother, while sometimes a boy will more closely resemble his mother. All these facts too may be regarded as evidence for my contention above, that both man and woman have male and female sperm.

9. Another thing which sometimes happens is that children are undersized and sickly, although both parents are large-bodied and strong. If this occurs subsequently to the birth of several children who are healthy like their parents, then it is clear that the child's sickliness began in the womb; the womb was more open than normal, and some of the child's nutriment from the mother escaped, hence the weakness of the child – and of course all living things fall sick to a degree in proportion to their normal strength. If on the other hand *all* the children born to a particular mother are weakly, the cause is the constriction of the womb. For if the space in which the embryo is nurtured is not adequate, obviously it will be under-sized, since it will have insufficient space to grow in. Whereas if (a) it has plenty of space and (b) it contracts no sickness, then it is reasonable to expect that a large offspring will be born to large parents. It is similar to what happens if you place in a narrow vessel a cucumber which has finished flowering but is still young and still growing from the bed. The cucumber will grow to a size and shape equal to the inside of the

vessel. But if you place it in a large vessel – one which is large enough to take a cucumber but which does not greatly exceed the natural size of the plant – then the cucumber will grow to a size and shape equal to the interior of this vessel: in its growth it attempts to rival the space in which it grows. In fact, it is generally true that all plants will grow in the way one compels them to. It is the same with the child: if he has plenty of space during his period of growth, he becomes larger; whereas if the space is confined, he will be smaller.

10. When the child is deformed in the womb, I consider that this occurs (a) as the result of a contusion. The mother has received a blow in the part where the embryo is, or has had a fall, or suffered some other violence. A deformity results in the place where the contusion occurred. When the contusion is extensive, the membrane enveloping the embryo is broken, and it is aborted. (b) Children may be deformed in another way: if there is some constriction in that region of the womb which is contiguous to the part in which the embryo is deformed, it must be the case that deformity occurs there as a result of the embryo's movement in the constricted space. A similar thing happens to trees which have insufficient space in the earth, being obstructed by a stone or the like. They grow up twisted, or thick in some places and slender in others, and this is what happens to the child as well, if one part of the womb constricts some part of its body more than another.

11. The children of deformed parents are usually sound. This is because although an animal may be deformed, it still has exactly the same *components* as what is sound. But when there is some disease involved, and the four innate species of the fluid from which the seed is derived form sperm which is not complete, but deficient in the deformed part, it is not in my opinion anomalous that the child should be deformed similarly to its parent.

So much then for this subject: I shall now return to my main argument.

THE NATURE OF THE CHILD

12.* If the seed which comes from both parents remains in the womb of the woman, it is first of all thoroughly mixed together – for the woman of course does not remain still – and gathers into one mass which condenses as the result of heat. Next, it acquires breath, since it is in a warm environment. When it is filled with breath, the breath makes a passage for itself in the middle of the seed and escapes. Once this passage of escape for the warm breath has been formed, the seed inspires from the mother a second quantity of breath, which is cool. It continues to do this throughout the whole period: the warmth of its environment heats it, and it acquires cold breath from the mother's breathing. In fact everything that is heated acquires breath: the breath breaks a passage for its escape to the outside, and through this break the object which is being heated draws a second lot of cold breath, by which it is fed. The same process occurs with wood, or with leaves, or with food and drink, when they are heated vigorously. You can see what happens from the case of burning wood – any kind of wood will behave in the same way, but green wood in particular. It will expel air where it has been cut, and when this air gets outside, it eddies around the cut. This is a matter of common observation, and the inference is obvious: the air in the wood, since it is hot, draws to itself cold air to feed upon, at the same time as it expels air. If this were not the case, then neither would the air eddy as it is expelled. For everything which is heated is fed by a proportionate quantity of cold. Now when the fluid in the wood is heated, it becomes air which then passes outside. As this air is expelled, the heat in the wood draws in cold air to replace it and to nourish itself. Green leaves also do the same when they are burned, for they contain air; and this air breaks a passage out for itself and escapes, eddying as it goes and making a crackling sound at the place where the air is also drawn in. Legumes, cereals and nuts also form air when they are heated, and this air makes a fissure

---

*In modern editions the numbering of the chapters of *The Nature of the Child* follows on from that of *The Seed*.

and escapes, and if these materials are moist, the greater the
quantity of air released, and the larger the fissure. However,
there is no need to labour the point that everything which is
subjected to heat both emits air and draws in the nutriment of
cold air by the same passage. Such then are the proofs which I
adduce in support of the contention that the seed, heated in
the womb, both contains and emits breath. However, there is
a second source of breath for the seed: this is the breathing of
the mother; for when the mother breathes in cold air from
the outside, the seed gets the benefit of it: the seed is made
warm by the warmth of its environment, and so it contains and
emits breath.

As it inflates, the seed forms a membrane around itself; for
its surface, because of its viscosity, stretches around it without a break, in just the same way as a thin membrane is formed
on the surface of bread when it is being baked; the bread rises
as it grows warm and inflates, and as it is inflated, so the
membraneous surface forms. In the case of the seed, as it
becomes heated and inflated the membrane forms over the
whole of its surface, but the surface is perforated in the middle
to allow the entrance and exit of air. In this part of the membrane there is a small projection, where the amount of seed
inside is very small; apart from this projection the seed in its
membrane is spherical.

13. As a matter of fact I myself have seen an embryo which
was aborted after remaining in the womb for six days. It is
upon its nature, as I observed it then, that I base the rest of my
inferences. It was in the following way that I came to see a
six-day-old embryo. A kinswoman of mine owned a very
valuable danseuse, whom she employed as a prostitute. It was
important that this girl should not become pregnant and thereby lose her value. Now this girl had heard the sort of thing
women say to each other – that when a women is going to
conceive, the seed remains inside her and does not fall out.
She digested this information, and kept a watch. One day she
noticed that the seed had not come out again. She told her
mistress, and the story came to me. When I heard it, I told
her to jump up and down, touching her buttocks with her

heels at each leap. After she had done this no more than seven times, there was a noise, the seed fell out on the ground, and the girl looked at it in great surprise. It looked like this: it was as though someone had removed the shell from a raw egg, so that the fluid inside showed through the inner membrane – a reasonably good description of its appearance. It was round, and red; and within the membrane could be seen thick white fibres, surrounded by a thick red serum; while on the outer surface of the membrane were clots of blood. In the middle of the membrane was a small projection: it looked to me like an umbilicus, and I considered that it was through this that the embryo first breathed in and out. From it, the membrane stretched all around the seed. Such then was the six-day embryo that I saw, and a little further on I intend to describe a second observation which will give a clear insight into the subject. It will also serve as evidence for the truth of my whole argument – so far as is humanly possible in such a matter. So much then for this subject.

14. The seed, then, is contained in a membrane, and it breathes in and out. Moreover, it grows because of its mother's blood, which descends to the womb. For once a woman conceives, she ceases to menstruate – except in some cases where a very small amount appears, no more than a token, during the first month – otherwise the child will be unhealthy. The blood instead descends from the whole body of the woman and surrounds the membrane on the outside. This blood is drawn into the membrane along with the breath, where the membrane is perforated and projects; and by coagulating, it causes the increase of what is to become a living thing. In due course, several other thin membranes form within the original one, these being formed in the same way as the first. Like it, they too extend from the umbilicus, and there are connections between them.

15. At this stage, with the descent and coagulation of the mother's blood, flesh begins to be formed, with the umbilicus, through which the embryo breathes and grows, projecting from the centre.

The reason that a pregnant woman does not suffer from the

fact that the menses have ceased to flow, is that the blood is no longer agitated by a massive flux occurring once a month. Instead, it flows gently into the womb in small amounts each day without causing discomfort, and what is contained in the womb increases. And it flows each day, rather than all at once every month, because the embryo in the womb draws it continually from the body, in proportion to its strength. Its respiration also works in the same way: respiration is at first slight, and the amount of blood flowing from the mother is slight as well. As respiration increases, the embryo draws the blood to itself more vigorously, and the amount flowing into the womb increases in quantity.

The reason why women who are not pregnant experience pain when the menses do not flow is this: in the first place, the blood in the body is set in agitation each month, by the following cause. There is a great difference in temperature between month and month. Now a woman's body has more fluid than that of a man, and is therefore sensitive to this change; and as a result her blood becomes agitated, fills up her veins and flows away from her. This is simply a fact of woman's original constitution. The result of all this is that when a woman is emptied of blood, she conceives; whereas if she is full of blood, she does not conceive. It is when the womb and veins are empty that women conceive; hence the most favourable time for conception is just after menstruation, and the reason is as I have stated. Now when the blood is agitated and secreted but, instead of flowing away, flows into the womb and the womb does not release it, then the womb is heated by the blood which lingers in it and in turn heats the rest of the body. Sometimes too the womb will discharge blood into the veins, and these, becoming filled, grow painful and cause swellings. There is sometimes a danger that a woman will be actually crippled when this happens. Sometimes the womb settles against the bladder and, by pressing upon it and closing it, causes a strangury; or sometimes it settles against the hip or the lumbar regions, causing pain there. In some cases, the blood has been known to remain in the womb for as long as five or six months, where it corrupts and becomes pus. Some women pass out this pus through the vagina, while in other cases it forms a tumour in the groin, so that the pus is expelled in that way. In fact, women suffer many maladies of this kind when their menses do not flow, but this is not the proper place to discuss them: they will be described in my course on women's diseases. I will now pick up my argument where I left it off.

16. Once the flesh is formed, the membranes continue to grow commensurately with the embryo. These form pouches, particularly the outside membrane. Into these pouches passes whatever blood is left over and serves no useful purpose after the embryo has drawn it from the mother by respiration and used it to grow with. When these membranes form pouches and fill with blood, they are called the 'chorion'.

17. As the flesh grows it is formed into distinct members by breath. Each thing in it goes to its similar – the dense to the dense, the rare to the rare, and the fluid to the fluid. Each settles in its appropriate place, corresponding to the part from which it came and to which it is akin. I mean that those parts which came from a dense part in the parent body are themselves dense, while those from a fluid part are fluid, and so with all the other parts: they all obey the same formula in the process of growth. The bones grow hard as a result of the coagulating action of heat; moreover they send out branches like a tree. Both the internal and external parts of the body now become more distinctly articulated. The head begins to project from the shoulders, and the upper and lower arms from the sides. The legs separate from each other, and the sinews spring up around the joints. The mouth opens up. The nose and ears project from the flesh and become perforated; while the eyes are filled with a clear fluid. The sex of the genitals becomes plain. The entrails too are formed into distinct parts. Moreover, the upper portions of the body now respire through the mouth and nostrils, with the result that the belly is inflated and the intestines, inflated from above, cut off respiration through the umbilicus and put an end to it. A passage outside is formed from the belly and intestines through the anus, and another one through the bladder.

Now the formation of each of these parts occurs through respiration – that is to say, they become filled with air and separate, according to their various affinities. Suppose you were to tie a bladder onto the end of a pipe, and insert through the pipe earth, sand and fine filings of lead. Now pour in water, and blow through the pipe. First of all the ingredients will be thoroughly mixed up with the water, but after you have blown for a time, the lead will move towards the lead, the sand towards the sand, and the earth towards the earth. Now allow the ingredients to dry out and examine them by cutting around the bladder: you will find that like ingredients have gone to join like. Now the seed, or rather the flesh, is separated into members by precisely the same process, with like going to join like. So much, then, on that subject.

18. By now the foetus is formed. This stage is reached, for the female foetus, in forty-two days at maximum, and for the male, in thirty days at maximum. This is the period for articulation in most cases, take or give a little. And the lochial discharge too after birth is usually completed within forty-two days if the child is a girl. At least this is the longest period which completes it, but it would still be safe even if it took only twenty-five days. If the child is a boy, the discharge takes thirty days – again the longest period, but there is no danger even if it takes only twenty days. During the latter part of the period the amount which flows is very small. In young women, the discharge takes a smaller number of days; more, when women are older. It is the women who are having their first child who suffer the most pain during the birth and during the subsequent discharge, and those who have had fewer children suffer more than those who have had a greater number.

Now then, the reason for the discharge after birth is this: during the earlier period of gestation – up to forty-two days for a girl, and thirty for a boy – the least amount of blood flows to cause the embryo to grow, while after this period the amount increases right up to the time the woman gives birth. We must expect then that the lochial discharge will correspond and flow out in accordance with the number of these days.

The woman's birth-pangs begin in the following way: her

blood is agitated and becomes very warm through the movement of the infant who is sturdy by this time. Once the blood is agitated, there issues first of all (after the child) a thick bloody serum: this opens up a way for the blood, as in the case of 'the water on the table'.* This purgation then flows each day, up to the time I have stated; the initial amount being one and a half Attic measures,† more or less, and so on proportionately until the flow ceases. If the woman is in a good state of health and is likely to remain so, the blood flows as it does from a sacrificial animal, and clots quickly. But if her health is bad, the flow is less, unhealthy in appearance, and is slow to coagulate. (In such cases, if the pregnant woman has a disease which is not constitutional, she will die during the lochial period. If the lochial flow does not begin immediately during the first period of days, whether she is in good health or not, but subsequently the flow comes in a rush, whether induced by medication or spontaneously, the flow will be in proportion to the number of days during which it did not flow all at once.‡ For if a woman is not thoroughly purged of the lochia, she becomes gravely ill with a risk of dying, unless she is treated swiftly and the purgation is eventually induced.) The reason that I have introduced these details, is to show that the limbs are differentiated at the latest, in the case of girls, in forty-two days, and thirty days for boys; and I take as evidence for the assertion the fact that the lochial discharge lasts for forty-two days after a girl, and for thirty after a boy, these being the maximum periods.

And now I shall state the whole thing over again, for the sake of clarity. I maintain that the flow is proportionate, because while the seed is in the womb, the amount of blood which flows into the womb from the mother who carries a female child is least during the initial period of forty-two days, the period within which the limbs are differentiated. From

---

*Presumably the author refers to a familiar demonstration, either of air pressure or of surface tension, but nothing is known of its precise form.

† The Attic measure (Kotyle) was 0.226 litres.

‡ i.e. the lochia are identical with menstrual blood, which does not flow 'all at once' but day by day once conception has occurred. Cf. ch. 15.

this time, the flow increases. And again, in the case of a male child, the same thing happens, only this time the period is thirty days. Here is further evidence for the truth of these facts: in the days immediately following the receipt of seed in the womb, the quantity of blood flowing from the mother into the womb is least, and subsequently it increases. For if a large amount were to flow all at once, the embryo could not breathe, but would be choked by the large flow of blood. In the purgation, on the other hand, the proportion is inverse; for here the flow is greatest in the initial period, and it then decreases until it finishes. Again, many women have miscarried with a male child a little earlier than thirty days, and the embryo has been observed to be without limbs; whereas those that were miscarried at a later time, or on the thirtieth day, were clearly articulated. So too in the case of female embryos which are miscarried, the corresponding period being forty-two days, articulation is observed. Hence it is manifest, both by reasoning and by necessity, that the period of articulation is, for a girl, forty-two days, and for a boy, thirty. The evidence is to be found both in abortion and in lochial purgation; while the cause is that the female embryo coagulates and is differentiated later, since the female seed is both weaker and more fluid than the male; so that it necessarily follows in accordance with this explanation that the female coagulates later than the male. This is also why purgation takes longer after a female birth than after a male birth. I now return to the point where I left off.

19. Once the embryo's limbs are articulated and shaped, as it grows its bones become both harder and hollow, this too being effected by breath. Once the bones are hollow, they absorb the richest part of clotted blood from the flesh. In due course the bones at their extremities branch out, just as in a tree it is the tips of the branches which are last to shoot forth twigs. It is in the same way that the child's fingers and toes become differentiated. Further, nails grow on these extremities, because all the veins in the human body terminate in the fingers and toes. Now of these veins, the largest are those which are in the head, and next to these, those in the legs and in the upper

and lower arms. But it is in the feet and hands that both the veins and nerves are thinnest and most numerous and most dense, and the bones too. This is particularly true of the fingers and toes. Now it is because the fingers and toes have bones, veins and nerves which are small and dense that the nails, which are also thin and dense, grow out of them. Their effect is to cut short the extremities of the veins, preventing them from growing any further, or one from being longer than another. So that it is perfectly natural that the nails, being at the body's extremities, should be very dense, for what they come from is very dense.

20. The hair takes root in the head at the same time as the nails grow. Hair grows in the following way: it grows longest and most abundantly where the epidermis is most porous and where the hair can receive the right amount of moisture for its nutrition. Where the epidermis becomes porous subsequently, there too the hair grows subsequently – namely, on the chin, the pubes, and other such places. Both the flesh and the epidermis become porous at the time when sperm first makes its appearance; and at the same time the veins open out more than they did previously. For during childhood, the veins are narrow, and do not give passage to the sperm. The same is true of menstruation in girls: in their case, the way is opened for menstruation and the passage of sperm at the same time. Now when the epidermis becomes porous, pubic hair begins to grow both in girls and boys, the hair now having sufficient moisture for its nutriment. The explanation of the growth of hair upon a man's chin is the same: here too the epidermis becomes porous, when moisture coming from the head flows into it. For both during intercourse and during the intervening times, the hair has the proper amount of moisture for its nutriment; but this is so most of all when the fluid in its course from the head during intercourse is delayed by its arrival at the chin, which projects forward of the breast. The evidence that hair grows where the epidermis is most porous is this: if you were to burn the epidermis just enough to raise a blister, which you then healed, the epidermis on the scar would become dense and would not grow hair.

The reason that those who are made eunuchs while they are still children neither become pubescent nor grow hair on their chins, but are hairless over their whole body, is that no passage is opened for the sperm, and therefore the epidermis does not become porous anywhere on the body (I have stated earlier that the passage for the sperm is cut off in eunuchs). Women also are hairless on their chin and over the body, because during intercourse their bodily fluid is not agitated to the same extent as in men, and therefore does not make the epidermis porous.

Those who are bald are so because their constitution is phlegmatic: for during intercourse the phlegm in their heads is agitated and heated, and impinging upon the epidermis burns the roots of their hair, so that the hair falls out. For the same reason, eunuchs do not become bald, because they do not experience the violent movement of intercourse which would heat the phlegm and cause it to burn the hair-roots.

The explanation of grey hair is that after fluid has been agitated in the body over a long period, its whitest part is separated off and arrives at the epidermis: the hair grows whiter because the moisture which it attracts is whiter than it was previously, while the epidermis is whiter in places where the hair is white. Those too who have a white head of hair from birth show a whiter epidermis where the hair is white; for it is there that the whitest part of the bodily fluid is. In fact, the complexion of the skin and the colour of the hair correspond to the colour of the moisture which the flesh attracts – white, or red, or black. Having said so much on this subject, I return to the remainder of my discourse.

21. The embryo starts to move once the extremities of the body have branched and the nails and hair have taken root. The time which it takes for this to happen is three months for males, and four for females. That at least is generally the case, although some infants start to move earlier. The reason why a male embryo starts to move earlier is its greater strength; moreover, the male is compacted earlier, since the seed from which it comes is stronger and thicker. As soon as the embryo has started to move, the mother's milk makes its appearance:

her breasts swell and the nipples grow erect, although so far
the milk does not flow. The appearance and the flow of milk
occur later in those women whose flesh is of a dense texture,
and earlier in those who have loose-textured flesh. The cause
of lactation is as follows: when the womb becomes swollen
because of the child it presses against the woman's stomach,
and if this pressure occurs while the stomach is full, the fatty
parts of the food and drink are squeezed out into the omentum
and the flesh. The process is the same as when you smear a hide
with large quantities of oil and, after giving the hide time to
absorb the oil, you squeeze it, and the oil oozes out again
under pressure. In exactly the same way, the stomach, containing the fatty portions of food and drink, percolates the fat
into the omentum and the flesh, under the pressure of the
womb. If the woman's flesh has a loose texture, she feels the
effect of the percolation all the sooner; but later, if her flesh
is not of this type. Moreover, pregnant animals, provided they
are not diseased, grow fatter than animals which are not pregnant, although their food and drink is exactly the same. This
is also true of pregnant women. Now from this fatty substance,
which is warmed and white in colour, that portion which is
made sweet by the action of heat coming from the womb is
squeezed into the breasts. A small quantity goes to the womb as
well, through the same vessels: for the same vessels and others
similar to them extend alike to the breasts and the womb. When
it arrives in the womb it has the appearance of milk, and the
embryo uses a small quantity; while the breasts are filled with
it and swell. When the child is born, it is the act of suckling
itself which causes the milk to flow into the breasts, once the
whole process has been set going initially. For in fact, when
the breasts give suck the veins into them become more permeable, and because they become more permeable they can
attract the fatty substance from the stomach and pass it along
into the breasts. It is similar to the case of a man who enjoys
intercourse frequently: the veins become more permeable,
thereby inducing him to further intercourse.

22. Nutrition and growth depend on what arrives from the
mother into her womb; and the health or disease of the child

corresponds to that of the mother. In just the same way, plants growing in the earth receive their nutriment from the earth and the condition of the plant depends on the condition of the earth in which it grows. Now when a seed is planted in the earth, it is filled with moisture from it (the earth contains many different varieties of moisture, which is why it can nurture plants). Once the seed is filled with moisture, it becomes inflated and swells. Now there is a potency in the seed: when the lightest part of this potency is condensed and compressed by breath and the moisture in the seed, it turns into green shoots and breaks the seed open. This is what happens at first: the shoots sprout upwards, but once they have sprouted, then the moisture in the seed is no longer sufficient for their nutrition. So the seed and its shoots break open in a downward direction: the shoots force the seed to release downwards that part of its potency which has been left behind owing to its weight, and roots are produced extending from the shoots.

Once the plant has taken firm root below and begins to derive its nutriment from the earth, then the whole seed is absorbed by the plant and disappears, excepting the husk, which is the most solid part of the seed. Eventually this too rots in the earth and disappears. After a time, some of the shoots send out branches.

Now the plant, since it comes from a seed, that is, from something moist, while it remains tender and moist and strives to grow upwards and downwards, cannot put forth fruit. The reason is that its potency is not strong and rich enough to be compacted into fruit. But when time has made the plant firm and rooted it, it develops broad veins running upwards and downwards, so the substance it draws from the earth is no longer watery, but thicker and more fatty and greater in quantity. This substance is heated by the sun, and erupts into the tips of the plant, where it becomes fruit of the same kind as what it came from. The reason for the abundance of fruit, despite its small origin, is that every plant draws from the earth more potency than did the seed from which it originated, and this potency erupts not simply in one place but in many; and once the fruit has broken out, it is nurtured subsequently by the plant, which draws upon the earth and transmits what it draws to the fruit. It is the sun which ripens and firms the fruit, by evaporating its more watery part. So much then for those plants which grow out of seeds, from the earth and water.

23. When trees are grown from slips, on the other hand, what happens is this: the cutting has a fracture at its lower end (where it was broken off from the parent tree), the end which is placed in the earth. This end sends out roots in the following way: when the plant is in the earth, and draws up moisture from the earth, it swells and becomes full of breath, though this is not yet the case with the part projecting above the earth. But in the lower part, the breath and moisture cause the heaviest portion of the plant's potency to condense and break forth in a downward direction; from it grow delicate roots. Once the plant has taken root below, it draws moisture from the root and imparts it to the portion above the ground. And so in its turn the upper portion too swells and acquires breath; and such of the potency of the plant as is light condenses and grows out in the form of shoots: thereafter the plant grows both upwards and downwards. So the process of sprouting in the case of plants grown from seeds is just the opposite to the process in plants grown from slips: the shoots grow upwards from the seed before the roots grow downwards, whereas the tree takes root first, and only later puts out leaves. The reason is that there is a quantity of moisture in the seed itself, and since the seed is wholly contained in the ground, it has, initially, sufficient nutriment to feed its shoots until it becomes rooted. This is not the case with the cutting, for it does not grow from something else which provides the leaves with their initial nutriment; instead, the cutting is like the tree, which has its greatest bulk above the earth, and so cannot be filled with moisture above the earth without some great potency coming from below to transmit moisture to the upper parts. So that at first the slip must necessarily sustain itself from the earth by means of its roots, and only subsequently transmit upwards the moisture which it attracts from the earth, and so blossom into leaves, and grow.

24. I shall now describe what causes the plant to branch as

it grows. Once it has drawn an excessive amount of moisture from the earth, this superabundance causes it to break out at the point where most has collected and it is here that the plant branches. The reason that it increases in size both laterally and upwards and downwards is that the earth below its surface is warm in winter and cool in summer. This is because in winter the rain falling from the sky makes the earth moist and, since moisture is heavy, compressed upon itself. Accordingly, the earth becomes more dense and, because it has no large pores, it is unventilated. That is why the earth under its surface is warm in the winter. Dung too is warmer when it is compressed than if it is loosely packed, and in general, substances which are moist and compressed grow warm spontaneously and are quick to rot because they get burnt up by the heat: their compression prevents the air from penetrating them. Whereas if they are dry and loosely packed, they are much slower to grow warm and to rot. Corn and barley too, if they are moist and packed together, are more likely to grow warm than when they are dry and lie loosely. Leather garments also, if you tie them together and compress them very tightly, are consumed spontaneously as though by fire. This is something I have observed myself. In fact you only have to consider to realize that everything which is compressed upon itself grows warmer that what is loosely packed, the reason being that these substances cannot be ventilated and cooled by the wind. Now the case is the same with the earth: under its surface it is dense and compacted upon itself as a result of the weight of moisture in it, and so it is warm during winter since it has no ventilation for its heat. Now when rain falls upon the earth, and then sends forth an exhalation, this exhalation is obstructed by the earth's density, and is forced back into the water. This is why springs during the winter are both warmer and flow more abundantly than during the summer: the air exhaled from the water goes back into the water, since the earth is dense and the air cannot penetrate it. This large quantity of water breaks out and flows where it can, and makes a broader passage for itself than if it were only a small quantity. (For water in the earth does not remain at rest, but regularly flows downwards.) Whereas if it were the case that the earth during winter offered a passage through itself to the air exhaled by the water, the water flowing from the earth would be less in quantity, and springs would not be abundant during winter. I have mentioned all these facts to show that the earth under its surface is warmer during winter than during summer.

25. Now for the reason why during the summer the earth is colder under its surface than in winter. (i) During the summer the earth is porous and light, because the sun evaporates its moisture (the earth of course always has some moisture in itself, in smaller or larger quantities). (ii) All winds come from water – you can deduce the truth of this from the fact that winds always blow off rivers and clouds, and clouds are simply water cohering together in the air. So then: the earth is porous and light during the summer, and it contains water. This water flows downwards, and as it does so it exhales from itself a constant stream of air. This air permeates the earth, which is light and porous, and cools it; and the water itself is chilled at the same time. What happens is the same as if you were to compress tightly a skin containing water, and having made a breathing hole for the water with a needle point, or perhaps something a little larger, to hang the whole skin up and cause it to oscillate. You will find that no breath, but only water, passes through the perforation: the reason being that the water does not have a sufficiently wide passage to exhale air. Now this corresponds to water in the earth during winter. But if before hanging up the skin and causing it to oscillate you make a wide passage for the water, it will be the case that air passes through the perforation; for the air (which comes from the water as it moves) will have a sufficiently wide passage to escape through the skin, and that is why the air passes through the hole. This, then, is the case with the water in the earth during summer. It has a sufficiently wide passage, since the sun has made the earth porous by drawing up its water, and the air, which is cold because it comes from the water, penetrates the porous and light earth, and in this way makes the earth cool beneath its surface in summer. Furthermore the water, which is the cause of the coldness of the air in the

earth, receives this air back into itself as well as emitting it into the earth.

In the same way too, drawing water from a well keeps the air in motion like a puff of wind, and causes it to chill the water; whereas if the water is not drawn during the summer but is left standing, because of its density it does not admit the air from the earth to the same extent, nor does it emit it into the earth. Instead, when it is allowed to remain stagnant in the well and is not moved about, it is warmed by the sun and the air superficially at first, then the heat is transmitted down through successive layers. This is why well-water which is not drawn in the summer is warmer than that which is drawn (note that springs which come from deep underground are always cold in the summer). When in the winter water is drawn from the earth, it is initially warm – the earth itself being warm – but as time passes it becomes cold: this is obviously due to the effect of the air, which is cold, for the wind aerates the water, allowing breath to penetrate it. In the same way water which is drawn from a well during summer is cold initially, and then grows warm: the reason being that it is chilled by the air which circulates through the porous earth; whereas once it has been drawn and left standing for a while, it is observed to grow warm, because the air is warm. Water which is left undisturbed in a well during summer grows warm for the same reason. So much, then, on this subject.

26. To return, then, to my original point, that the earth below its surface is cool in summer, warm in winter, while above the surface the opposite is true. Now if a tree is to be sound, it must not be affected by two lots of heat simultaneously, nor two lots of cold: if it is affected by heat above the ground it must be affected by cold below the ground, and vice versa. The roots distribute to the tree whatever they attract, while the tree distributes whatever it attracts to the roots: there is in effect an equal dispensation of heat and cold. Just as when a man takes into his stomach those foods which cause heat when they are digested, he requires a compensatory cooling from drink, so too in the case of a tree the lower parts must compensate the upper, and vice versa. This is in fact the

reason why a tree grows both upwards and downwards: it is because it receives nutriment from above as well as from below. While it is still very tender it does not bear fruit: so far, it has no thick and fatty potency capable of producing fruit. But after a time, its veins become sufficiently wide to draw thick and fatty substance from the ground. The sun then melts this substance, lightening it, and causing it to erupt into the tree's extremities and become fruit. The sun also evaporates the thin portion of the moisture from the fruit, while it sweetens the thicker part by warming and ripening it. The reason why some kinds of tree do not bear fruit is that they contain insufficient fatty substance to contribute to fruiting.

Trees cease to grow once time has made them solid, and they have taken firm hold below with their roots. Some trees, however, grow from grafts implanted into other trees: they live implanted in these trees and produce fruit, but the fruit is different from that of the tree on which they are grafted. This is how: first of all the graft produces buds, for initially it still contains nutriment from its parent tree, and only subsequently from the tree in which it was engrafted. Then, when it buds, it puts forth slender roots into the tree, and feeds initially on the moisture actually in the tree in which it is engrafted. Then in course of time it extends its roots directly into the earth, through the tree in which it was engrafted; thereafter it uses the moisture which it draws up from the ground, and that is how it is nurtured – from the ground. There is therefore nothing anomalous in the fact that grafts bear different fruit: it is because they live from the ground. So much then for trees and their fruit – I could hardly avoid giving a complete account of the subject.

27. But to return to the main argument which was my reason for introducing these matters. I maintain, then, that all plants which grow in the ground live off the moisture which comes from the ground, and that the character of the plant depends on the character of this moisture. Now it is in just the same way that the child in the womb lives from its mother, and it is on the condition of health of the mother that the condition of health of the child depends. But in fact, if you review what

I have said, you will find that from beginning to end the process of growth in plants and in humans is exactly the same. So much for that subject.

28. The child while it is in the womb has its hands tucked against its chin, while its head lies near its feet. However it is not possible to decide with any accuracy whether its head is above or below – not even if you actually see the child in the womb. The child is held in its place by the membranes which extend from the umbilicus.

29. Now I come to the observation which I promised to describe a little earlier – one which will make the matter as clear as is humanly possible to anyone who wishes to know that the seed is contained in a membrane which has an umbilicus in the centre, and that the seed initially draws breath into itself and expires it, and that membranes extend from the umbilicus. Furthermore (if you accept the evidence which I am about to give) you will find that the growth of the infant is from beginning to end exactly as I have described it in my discourse. If you take twenty or more eggs and place them to hatch under two or more fowls, and on each day, starting from the second right up until the day on which the egg is hatched, you take one egg, break it open and examine it, you will find that everything is as I have described – making allowance of course for the degree to which one can compare the growth of a chicken with that of a human being. You will find for instance that there are membranes extending from the umbilicus – in fact, that in every point all the phenomena I have described in the human child are to be found in a chicken's egg also. Yet if a man had not actually seen it, he would find it hard to believe that there is an umbilicus in a chicken's egg. But it is so; so much, then, for that.

30. When it is time for the mother to give birth, what happens is that the child by the spasmodic movements of its hands and feet breaks one of the internal membranes. Once one is broken, then the others of course are weaker, and these break too in order of their proximity to the first, right up to the last one. When the membranes are broken, the embryo is released from its bonds and emerges from the womb all

bunched together. For nothing has any strength to hold it once the membranes fail, and when the membranes have carried away, the womb itself cannot hold the child back – in fact even the membranes in which the child is enwrapped are not fastened very strongly to the womb. Once the child is on its way, it forces a wide passage for itself through the womb, since the womb is resilient. It advances head first – that is the natural position, since its weight measured from above the navel is greater than it is below.

It is in its tenth month in the womb that it acquires an access of force sufficient to rupture the membranes, and that is when the mother gives birth. If however the child suffers from some violent injury, the membranes are ruptured and it emerges earlier than the appointed time. Another case in which the mother may give birth to her child earlier than the tenth month is when the nutriment coming from the mother to the child gives out sooner than that time. But those women who imagine that they have been pregnant longer than ten months – a thing I have heard them say more than once – are quite mistaken. This is how their mistake arises: it can happen that the womb becomes inflated and swells as the result of flatulence from the stomach, and the woman of course thinks that she is pregnant. And if besides her menses do not appear but collect in the womb, and there is a continuous flow into the womb (accompanied sometimes by the gas from the stomach while sometimes the menstrual blood grows heated as well),* then she is especially likely to imagine she is pregnant. After all, her menstruation has ceased and her belly is swollen. Then it sometimes happens that the menses break forth either spontaneously or because a further flow into the womb carries down with it what was already there; the wind is discharged; and in many cases immediately after the loss of the menses the womb gapes and is turned down towards the vagina. Now if they have intercourse with their husbands then, they conceive on the same day or a few days afterwards. Women who are inexperienced in these facts and their reasons then reckon their pregnancy to include the time when their menses did not

* The text and meaning of this passage are doubtful.

flow and their wombs were swollen. But in fact it is impossible for pregnancy to last longer than ten months, and I shall explain why.

The nutriment for growth which the mother's body provides is no longer sufficient for the child after ten months are up and it is fully grown. It is nurtured by drawing the sweetest part of the blood towards itself, although it is fed to some extent from the milk as well. Once these are no longer sufficient and the child is already big, in its desire for more nutriment than is there it tosses about and so ruptures the membranes. This occurs more frequently in women who are bearing their first child: with them, the supply of nutriment for the child tends to give out before the ten months are up. This is the reason: the menstrual flow of some women is sufficiently abundant, while with other women the flow is less. (If this is always the case, it is the result of the constitution which she has inherited from her mother.) Now it is the women whose menses are small in quantity who also provide their infants with insufficient nutriment towards the end of their term when the infant is already large, and so cause it to toss about and bring on birth before ten months are up. The reason is their small flow of blood. Usually too these women cannot give milk: this is because they have a dry constitution and their flesh is densely packed. My assertion then is that what brings on birth is a failure in food supply (excepting cases of actual injury). My evidence for this is as follows: consider the way in which a chicken develops from the yolk of the egg. The egg is made warm by the sitting of the hen, and its content is set in motion by the same cause. Now through being heated the content of the egg acquires breath, and then a second quantity of cold breath from the surrounding air, through the shell – the egg-shell is porous, allowing sufficient air to be drawn through it for what is contained in the egg. The chicken grows inside the egg and becomes articulated in approximately the same way as the child, as I have already said. Now the chicken itself originates from the yolk, but it gets its nutriment and increase from the white – a fact which is quite obvious to anyone who has studied the matter. When the supply of nutriment from the egg gives out, then not having enough to live on, it looks for more and moves about violently in the shell, and ruptures the membranes. When the hen notices the violent motion of the chick, she pecks at the shell and hatches it: this is on the twentieth day. It is clear that this is what happens, because when the hen breaks open the egg-shell there is practically no fluid left inside: it has all been used up by the chicken. It is the same with the child: once it has reached a certain size and its mother can no longer provide enough nutriment, in its search for more nutriment the infant tosses about until it ruptures the membranes, and being released in this way it emerges all at once. This occurs within a maximum of ten months. By the same principle animals both domestic and wild give birth at the proper time for each species, and no later: for there must necessarily be a definite time for every species of animal, at which the food supply becomes insufficient for the embryo and then gives out, and the animal gives birth. Those which have a smaller amount of nutriment for their embryos give birth earlier, while those which have a larger amount, later. So much, then, on that subject.

Once the membranes are ruptured, if the infant's momentum is in the direction of the head, the birth is easy for the mother. But if it comes sideways or feet first (this happens if its momentum inclines it in this direction, either because the size of the womb has given it space to move or because the mother has not kept still at the beginning of her birth-pangs) the birth is difficult and often fatal, either to the mother or to the child or both. In childbirth it is the women who are having their first child who suffer the most, because they have had no experience of the pain; apart from the general discomfort of the body, they suffer most in the loins and the hips, because these become distended. Those who have more experience of bearing children suffer less; much less, if they have had a large number of children.

If the embryo moves in the direction of the head, the head is first to emerge, followed by the limbs, with the umbilical cord, to which the chorion* is attached, coming last of all. After all these comes a bloody serum, which is secreted from the head and the rest of the body in consequence of violence, pain and heat. This serum opens up a way for the purging of the lochia: the lochia follow the serum and flow for the period already stated. The lochial discharge empties the veins, and in consequence of this the breasts collapse, along with those other parts of the body which contain much fluid. This occurs to the least extent after the birth of the first child, but subsequently to a greater extent, after more children have been born. So much, then, on that subject.

31. Twins are produced from one act of intercourse. The womb contains a number of crooked pouches, at varying distances from the vagina. Animals which produce large litters have a greater number of these pouches than those which give birth to small litters, and this is true of animals both domestic and wild, and birds. Now when it happens that the sperm on its arrival in the womb is divided into two pouches, neither of which releases it into the other, then each of these separate portions in each pouch forms a membrane and comes alive in just the same way as I have said the undivided seed does. The evidence that it is from one act of intercourse that twins are born is given by the dog, the pig, and other animals which produce two or more offspring from one act of intercourse; each separate embryo in the womb is contained in its pouch, with its own membrane – this is a matter of common observation – and these animals generally produce all their offspring on the one day. In the same way, when a woman has twins as the result of one intercourse, each is contained in its own pouch and has its own chorion, and she bears both of them on the same day, one emerging first, and then the other, each with its chorion.

As for the fact that twins are born of which one is male, the other female, I maintain that in every man and in every woman – in fact in every animal – there exist both weaker and stronger varieties of sperm. Now the sperm does not come all at once:

*For the author's definition of *chorion*, see chapter 16.

it comes out in two or three successive spasms. It is not possible that the first and the last lot should always be of even strength. The pouch which receives thicker and stronger sperm will contain a male, while that which receives sperm which is more fluid and weaker will contain a female. If strong sperm enters both, both will contain male offspring; if the sperm is weak, then both will contain female offspring.

With that I bring my whole account to its end.

THE HEART*

This short treatise is almost certainly a good deal later than most of the other works in the Hippocratic Corpus, the most commonly accepted view being that it belongs to a period approximately contemporary with, or slightly later than, Erasistratus of Ceos, who was active around 260 B.C. It is included in this selection as the outstanding work dealing with an anatomical subject in the Hippocratic Collection. It is the first extant treatise to mention the valves of the heart.

1. In shape the heart is like a pyramid, in colour a deep crimson. It is enveloped in a smooth membrane. In this membrane there is a small quantity of fluid, rather like urine, giving one the impression that the heart moves in a kind of bladder. The purpose of the fluid is to protect the pulsation of the heart, but there is just about sufficient of it to alleviate the heat of the heart as well. The heart filters out this fluid after it has received it and made use of it, drinking it up from the lung.

2. Now when a man drinks, most of it goes to the belly (the gullet being a sort of funnel which catches the greatest part of all that we ingest) but some of it goes into the larynx as well. The amount is, however, very small – not sufficient for us to feel it forcing its way in through the pressure of the current; for the epiglottis is a precisely fitting lid, and will not allow anything more than drink to get past. Here is the proof: take some water, colour it with blue copper carbonate or red ochre and give it to an animal which is almost dying of thirst (a pig is best, since this animal is neither cautious in its feeding habits nor refined in its manners). While the animal is still drinking, cut its throat.† You will find that it is stained with

---

*The printed texts of *de Corde* are all more or less unsatisfactory. For the most part the text of F. C. Unger has been used (*Mnemosyne*, 51 (1923), pp. 5off.): this differs considerably from that of Littré. The translation and interpretation of some passages have been discussed in I. M. Lonie, *Medical History*, vol. 17 (1973), pp. 1–15, 136–53.

† Presumably the author means the windpipe.

what the animal has drunk. The operation however is one which requires uncommon skill. We need have no doubts then when we say that drink has an effect, and a beneficial one, upon the larynx. Why is it then that when water rushes in impetuously it inconveniences us and provokes an acute fit of coughing? Because, I say, its course is then contrary to that of the breath; but the liquid which flows in as a result of pressure passes down the side of the larynx and hence does not oppose the upward passage of breath. Quite the contrary, in fact: by moistening the larynx it provides a smooth passage for the air.

3. The heart draws this fluid from the lung along with the air. But while it must necessarily expel the air once this has performed its service, by the route by which it drew it in, it allows some of the fluid to dribble into its sheath, letting the rest pass out along with the air. In this way too the breath, when it runs back, raises the palate. Of course the breath must run back in due course, since it is not food – how could mere air and water be food for man? They are rather a compensation for an innate deficiency.

4. But to return to my subject, the heart. The heart is an exceedingly strong muscle – 'muscle' in the sense not of 'tendon' but of a compressed mass of flesh. It contains in one circumference two separate cavities, one here, the other there. These cavities are quite dissimilar: the one on the right side lies face downwards, fitting closely against the other. By 'right' I mean of course the right of the left side, since it is on the left side that the whole heart has its seat. Furthermore this chamber is very spacious, and much more hollow than the other. It does not extend to the extremity of the heart, but leaves the apex solid, being as it were stitched on outside.*

5. The other cavity lies somewhat lower, and extends towards the line of the left nipple, which in fact is where its pulsation is observed. It has a thick surrounding wall, and is hollowed out inside to a cavity like that of a mortar. It is enwrapped and cushioned in the lung, and being surrounded

---

*He visualizes the right ventricle as enfolding the left: see ch. 9.

by it, it controls and tempers its own heat. For the lung is both cold in itself and is also cooled by respiration.

6. The inside surface of both chambers is rough, as though slightly corroded; the left more so than the right, for the innate heat is not situated in the right. It is therefore quite to be expected that the left chamber should be rougher than the right, being filled as it is with untempered heat. Hence it is of such massive construction – to protect it against the strength of the heat.

7. The orifices of the cavities are not exposed until one cuts off the tops of the ears* and removes the head [i.e. the base] from the heart. Once they are cut off, a pair of orifices for the two chambers is revealed. For the thick vein, running up from one, escapes our eye unless we dissect.†

These are the springs of man's existence: from them spread throughout his body those rivers with which his mortal habitation is irrigated, those rivers which bring life to man as well, for if ever they dry up, then man dies.

8. Close by the origin of the blood-vessels certain soft and cavernous [or 'porous'] bodies enfold the heart. Although they are called 'ears' they are not perforated as ears are, nor do they hear any sound. They are in fact the instruments by which nature catches the air – the creation, as I believe, of an excellent craftsman, who seeing that the heart would be a solid thing owing to the density of its material, and in consequence would have no attractive power, he equipped it with bellows, as smiths do their furnaces, with which the heart controls its

---

*By 'ears' the author evidently means the auricles along with the atria, since he does not otherwise refer to the atria. See note on p. 353.

†The text of these two sentences is dubious and the meaning is ambiguous. It is not clear whether two orifices are meant, or four; and the remark about the 'thick vein' is obscure. The problem is discussed in I. M. Lonie, op. cit. Dissection reveals four orifices, which are the sites of the four valves of the heart. The 'thick vein' is the *vena cava superior* and *inferior*, and the author's point is that before dissection the *vena cava* appears to be two veins, which would suggest that there are *five* main vessels, not four, leading from the heart; whereas dissection shows, in the author's opinion, that the *vena cava* is really one vein, whose common stock he takes to be the right atrium.

respiration. Here is the evidence for my statement: you can see the heart pulsing in its entirety, while the ears have a separate movement of their own as they inflate and collapse.

9. It is for this reason too that I maintain that while inspiration into the left cavity is effected through the veins, in the right cavity it is effected through an artery; for vessels which are soft have more attractive power, being more capable of distension. Now there was a need for that part of the heart which enfolds it like a cloak to receive less cooling, for it is not on the right side that the heat is situated. Hence it received an organ which would be favourable to this defect, to prevent it from being entirely overcome by the air entering it.*

10. The rest of my account will be concerned with the hidden membranes of the heart – a piece of craftsmanship deserving description above all others. There are membranes in the cavities, and fibres as well, spread out like cobwebs through the chambers of the heart and surrounding the orifices on all sides and emplanting filaments into the solid wall of the heart.† In my opinion these serve as the guy-ropes and stays of the heart and its vessels, and as foundation to the arteries. Now there is a pair of these arteries, and on the entrance of each three membranes have been contrived, with their edges rounded to the approximate extent of a semicircle. When they come together it is wonderful to see how precisely they close off the entrance to the arteries. And if someone who fully understands their original arrangement removes the heart from a cadaver and while propping up one membrane he leans the other against it he will find that neither water nor

---

*This whole passage has been convincingly emended in Unger's text which, with one alteration, is what has been translated. What the author somewhat obscurely means is that the heart draws air from the lungs into both ventricles, into the left through the pulmonary vein, and into the right through the pulmonary artery. But the right ventricle, having no 'innate heat' of its own, requires only a very small amount of air: hence it is equipped with the thicker-walled artery (evidently the semilunar valves on the pulmonary artery, which in chapter 10 are said to fit less precisely than those on the aorta, admit the passage of air).

† The author seems to recognize the *chordae tendineae* with the *musculi papillares*, and the *trabeculae carneae*.

air can be forced into the heart. This is especially true in the case of the membranes in the left chamber, which are engineered more precisely.

This is what one would expect: for man's intelligence, the principle which rules over the rest of the soul, is situated in the left chamber.

11. Its nutriment is neither the solid food nor the drink which comes from the belly, but a pure and luminous substance which is refined out of the blood. It conveys this nutriment out of the neighbouring blood receptacle by transmitting its rays, deriving it from there as though from the belly and intestines, which of course it does not do in reality. Now to avoid any disturbance from the confused movement of the food in the great artery [i.e. the aorta], it closes off the passage to that artery. For the great artery feeds from the belly and intestines and is laden with food not fit for the ruling principle.

It is obvious that the left cavity is not nourished by visible blood: if you kill an animal by cutting its throat and open up the left chamber, you will find it quite empty apart from some serum and bile, and the membranes which I have described. But the artery will not be empty of blood, nor the right chamber. This, then, as I see it, is the reason for the membranes on this vessel.

12. As for the vessel which comes from the right chamber, this too is closed off by the membranes meeting together. But they are too weak to close it completely. It opens towards the lung to provide blood for the lung's nutriment, while towards the heart it is closed, but not hermetically, so that air can enter it, though only in a very small quantity. For in this part the heat is weak, being dominated by a mixture of cold. Contrary to the general opinion, blood is in fact not a hot thing by nature, any more than any other fluid; though it may be made hot.

Such, then, is my description of the heart.

NOTE ON THE TRANSLATION OF SOME HIPPOCRATIC TERMS IN *THE NATURE OF THE CHILD* AND *THE HEART*

Phleps, plural *phlebes*. In Greek medical writers this word generally means 'blood-vessel', which includes both the vein and the artery of modern anatomy. After the distinction had been made, *phleps* approximated to the sense in which the modern anatomist uses 'vein'. Yet even the author of *The Heart*, who is aware of the distinction, in one passage (ch. 8) uses *phleps* in its general sense, and I have translated 'blood-vessels' accordingly. There is no difficulty here. However, in *The Nature of the Child*, *phleps* is used for the author's peculiar hypothesis that the body is a network of communicating vessels which convey all sorts of fluid: sperm, milk, the 'humours', as well as blood. 'Blood-vessel' would be a misleading translation here, since it would carry the implication that the physiological processes described are abnormal. In fact *phleps* for this author simply means any vessel in the body which conveys fluid. I have therefore preferred the translation 'vein' to 'blood-vessel', which has the further advantage that 'vein' may also be applied to porosities in other substances, e.g. the 'veins' in plants, and in fact the author of this treatise uses *phleps* in this way.

Neuron, plural *neura*. There is a similar difficulty with this word, which cannot simply be translated as 'nerve,' since before the discovery of the nervous system the word was much more likely to signify what we should call tendons and ligaments. Yet it may also refer, anatomically if not physiologically, to what the modern anatomist would recognize as nerves. Since English has no word which covers all three meanings, it is not possible to give a consistent translation here. Hence the same word is translated as 'ligaments' in chapter 2 of *The*

Seed and as 'nerves' in chapter 19 of *The Nature of the Child.*

Ous, plural *ōta*, and *gastēr*, plural *gasteres*. Literally 'ear' and 'belly'. In *The Heart* it is tempting to translate these as 'auricle' and 'ventricle' respectively, since the words contain the metaphor from which the modern terminology originated. But 'auricle' would be definitely misleading, since the author clearly includes in the 'ears' of the heart the atria, while 'ventricle' implies the modern distinction between ventricle and atrium. 'Cavity' seemed a safer translation here.

Dynamis, plural *dynameis*. This word, which signifies a concept or range of concepts integral to Greek science, is notoriously difficult to translate. In general it means 'power', but also 'quality', which in early Greek thought is not always distinguished from the thing or substance which possesses that quality. The author of *The Nature of the Child* abnormally extends this tendency when he uses *dynamis* to signify 'substance of a particularly active character'. I have attempted to preserve this semantic anomaly by using the word 'potency': it should be clear from the context that the author means a substance. The closest translation would be the word 'virtue' in its archaic use: Chaucer's

> And bathed every veyne in swich licour
> Of which vertu engendred is the flour*

conveys the author's idea exactly. 'Vertu' is of course Latin *virtus*, which was the standard translation of Greek *dynamis*. The passage from Chaucer is particularly apt, since it equates 'licour' with 'vertu', just as the author of *The Nature of the Child* equates his 'potency' with the 'moisture' (*ikmas*) which is contained in the seed.

*Prologue to *The Canterbury Tales*, ll. 3–4.

BIBLIOGRAPHY

EDITIONS AND COMMENTARIES

The editions used for the translations in this volume are the following:

H. KUEHLEWEIN, *Hippocratis opera quae feruntur omnia*, Teubner edition, 2 vols., Leipzig, 1894–1902, for *Tradition in Medicine*, *Epidemics* I and III, *Airs, Waters, Places* and *Regimen in Acute Diseases*.

W. H. S. JONES, *Hippocrates*, Loeb edition, vols. 1, 2 and 4, London, 1923–31, for *The Oath, The Canon, The Science of Medicine, Prognosis, Aphorisms, The Sacred Disease, Dreams, The Nature of Man* and *A Regimen for Health*.

E. T. WITHINGTON, *Hippocrates*, Loeb edition, vol. 3, London, 1928, for *Fractures*.

I. M. LONIE's forthcoming Ars Medica (Berlin) edition of *The Seed* and *The Nature of the Child*.

F. C. UNGER's edition of *The Heart* from *Mnemosyne*, New Series, vol. 51, 1923, pp. 50–57.

The most recent complete edition of the Hippocratic Corpus is that of E. Littré, *Oeuvres complètes d'Hippocrate*, 10 vols., Paris, 1839–61.

Other important editions are those in the Corpus Medicorum Graecorum series (Teubner, Leipzig, now Akademie-Verlag, Berlin: three volumes have been published to date: I, 1, ed I. L. Heiberg, Leipzig, 1927; I, 1, 2, ed. H. Diller, Berlin, 1970, and I, 2, 1, ed. H. Grensemann, Berlin, 1968); in the Budé series (Les Belles Lettres, Paris: volumes VI,1, VI,2 and XI, all edited by R. Joly, have so far appeared); and in the Ars Medica series (de Gruyter, Berlin: volume 1, H. Grensemann's edition of *The Sacred Disease*, appeared in 1968).

The following other editions of *Tradition in Medicine* and *Prognosis* have particularly valuable commentaries:

W. H. S. JONES, 'Philosophy and Medicine in Ancient Greece' (Suppl. 8 to the *Bulletin of the History of Medicine*, Baltimore, 1946).

BIBLIOGRAPHY

A. J. FESTUGIÈRE, 'Hippocrate, L'Ancienne médecine', *Etudes et commentaires*, 4, Paris, 1948.

B. ALEXANDERSON, *Prognostikon (Studia Graeca et Latina Gothoburgensia*, 17), Göteborg, 1963.

SUGGESTIONS FOR FURTHER READING

The most important general works on Hippocratic medicine generally accessible in English are:

W. A. HEIDEL, *Hippocratic Medicine: Its Spirit and Method*, New York, 1941.

L. EDELSTEIN, *Ancient Medicine*, Baltimore, 1967 (this contains important papers on Hippocratic medicine, later Greek medicine and Greek science).

Other important recent studies are:

L. BOURGEY, *Observation et expérience chez les médecins de la collection hippocratique*, Paris, 1953.

J.-H. KÜHN, 'System- und Methodenprobleme im Corpus Hippocraticum', *Hermes Einzelschriften*, 11, Wiesbaden, 1956.

R. JOLY, *Le Niveau de la science hippocratique*, Paris, 1966.

F. KUDLIEN, *Der Beginn des medizinischen Denkens bei den Griechen*, Zürich and Stuttgart, 1967.

H. FLASHAR, (ed.), *Antike Medizin*, Darmstadt, 1971 (an important collection of specialized papers).

R. JOLY, 'Hippocrates of Cos', *Dictionary of Scientific Biography*, ed. C. C. Gillispie, New York, 1972, pp. 418–31.

J. JOUANNA, *Hippocrate: pour une archéologie de l'école de Cnide*, Paris, 1974.

G. E. R. LLOYD, 'The Hippocratic question', *Classical Quarterly*, 25, 1975, pp. 171–92.

H. GRENSEMANN, *Knidische Medizin*, I, Berlin and New York, 1975.

L. BOURGEY and J. JOUANNA, (eds.), *La Collection hippocratique*, Leiden, 1975.

Bourgey's and Kühn's books, and Joly's and Lloyd's articles, have full references to the most important earlier work, and Flashar's book has a comprehensive, up-to-date bibliography.

On Greek science in general, see:

G. SARTON, *A History of Science*, 2 vols., London, 1953–9.

BIBLIOGRAPHY

B. FARRINGTON, *Greek Science* (revised one-volume edition), Penguin, London, 1961.

G. E. R. LLOYD, *Early Greek Science, Thales to Aristotle*, London, 1970.
Greek Science after Aristotle, London, 1973.

The literature on Hippocrates' subsequent influence is immense. Apart from the classic histories of medicine, such as M. Neuburger, *History of Medicine* (trans. E. Playfair), 2 vols., London, 1910–25, and F. H. Garrison, *An Introduction to the History of Medicine*, 4th ed., Philadelphia and London, 1929, the following specialized papers provide a good introduction to the topics with which they deal:

O. TEMKIN, 'Geschichte des Hippokratismus im ausgehenden Altertum', *Kyklos*, 4, 1932, pp. 1–80.

P. O. KRISTELLER, 'The School of Salerno', *Bulletin of the History of Medicine*, 17, 1945, pp. 138–94.

P. KIBRE, 'Hippocratic Writings in the Middle Ages', *Bulletin of the History of Medicine*, 18, 1945, pp. 371–412.

O. TEMKIN, 'Die Krankheitsauffassung von Hippokrates und Sydenham in ihren "Epidemien"', *Archiv für Geschichte der Medizin*, 20, 1928, pp. 327–52.

GLOSSARY OF NAMES*

AETIUS OF AMIDA. Sixth-century A.D. author of a sixteen-book medical encyclopedia.

ALCMAEON OF CROTON. *Fl. c.* 450 B.C. A natural philosopher interested also in medicine. He has been thought, on the basis of a report in Chalcidius, to have been one of the first Greeks to have attempted dissection.

ALEXANDER OF TRALLES. Sixth-century A.D. medical encyclopedist. His treatise in twelve books is extant.

ANAXAGORAS OF CLAZOMENAE. *c.* 500–*c.* 428 B.C. A natural philosopher who worked mainly in Athens; teacher and friend of Pericles.

ARISTOTLE OF STAGIRA. 384–322 B.C. The most important of the extant works from the point of view of the medical sciences are the zoological treatises, especially *De Generatione Animalium*. He refers, in his psychological treatises, to a discussion *On Health and Disease*, though this work is not extant.

CAELIUS AURELIANUS. Fifth-century A.D. medical writer. His *Acute Diseases* (three books) and *Chronic Diseases* (five books) are thought to be largely derived from works of Soranus (q.v.).

CASSIODORUS. Sixth-century A.D. historian and (through his *Institutions*) influential educationalist.

CELSUS. First-century A.D. encyclopedist. His *On Medicine* (in eight books) is the only part of his six-part encyclopedia that is extant.

DEMOCEDES OF CROTON. *Fl. c.* 525 B.C. According to Herodotus (III,129 ff.) one of the most famous doctors of his day, being, at different times, physician to Polycrates of Samos and (while a prisoner in Susa) to Darius.

DEMOCRITUS OF ABDERA. *c.* 460–*c.* 370 B.C. According to Diogenes Laertius he composed several works on medical topics. He is chiefly famous for having developed the atomic theory originally suggested a little earlier by Leucippus. Since none of Democritus' treatises is extant, this has to be reconstructed from fragments and references in Aristotle and other sources.

EMPEDOCLES OF ACRAGAS. *Fl. c.* 445 B.C. Author of a cosmo-

* Including Greek and Latin medical writers important for the understanding of Hippocratic writings and their influence.

logical poem *On Nature* (which sets out the theory of the four physical elements, earth, water, air and fire) and of a religious poem the *Purifications*, fragments of both of which are extant. He is described by Galen as belonging to the Italian school of medicine. Yet references to the 'word of healing' in a quasi-mystical context in the *Purifications* suggest that his medicine may have been of the type criticized by the author of the Hippocratic work *The Sacred Disease*.

ERASISTRATUS OF CEOS. *Fl. c.* 260 B.C. Famous physician, anatomist and physiologist. His works (all lost) are frequently cited and criticized by Galen who ascribes to him (for example) the discovery of the valves of the heart. According to Celsus (q.v.) he and Herophilus practised dissection and vivisection on humans.

GALEN OF PERGAMUM. A.D. 129–*c.* 210. An outstanding physician, anatomist and physiologist, he also wrote extensively on non-medical subjects, for example logic. Many of his medical and biological treatises, including thirteen of his commentaries on Hippocratic works, are extant.

HEROPHILUS OF CHALCEDON. *Fl. c.* 270 B.C. Famous physician, anatomist and physiologist who (like Erasistratus, q.v.) is reported to have dissected and vivisected human subjects. He made many anatomical discoveries and developed the first systematic theory concerning the use of the pulse in diagnosis.

HIPPOCRATES OF COS. The traditional date of his birth is 460 B.C. He is mentioned by Plato, Aristotle and Meno (q.v.) but he cannot be identified with certainty as the author of any one of the Hippocratic treatises.

ISIDORE OF SEVILLE. *Fl. c.* A.D. 600. Historian and encyclopedist; among the extant works are the *On the Nature of Things* and the *Etymologies*, both of which allude briefly to medical matters.

MACROBIUS. Fifth-century A.D. author of the influential *Commentary on the Dream of Scipio* and the popular miscellany the *Saturnalia*.

MELISSUS OF SAMOS. *Fl. c.* 440 B.C. Followed Parmenides (q.v.) in denying change and plurality. Mentioned in the Hippocratic work *The Nature of Man*.

MENO. Pupil of Aristotle. He wrote a history of medicine at the end of the fourth century B.C., excerpts from which are preserved in the papyrus Anonymus Londinensis.

ORIBASIUS OF PERGAMUM. *Fl.* A.D. 360. Some twenty-five books of his seventy-book medical encyclopedia, the *Medical Collections*, largely based on Galen, are extant.

PARMENIDES OF ELEA. Philosopher born *c.* 515 B.C. His denial of change and plurality and his rejection of sensation in preference for reason influenced both later fifth-century B.C. natural philosophy and Plato profoundly.

PAUL OF AEGINA. Seventh-century A.D. medical encyclopedist. His general medical treatise in seven books is extant.

PHILISTION. Fourth-century B.C. medical writer whose theories are reported in Anonymus Londinensis. He is referred to in Plato's Second Letter and is thought to have influenced the medical doctrines that Plato included in the *Timaeus*. He is mentioned by Galen as one of the authors to whom the Hippocratic works *Regimen* and *A Regimen for Health* were attributed.

PHILOLAUS OF CROTON. *Fl. c.* 410 B.C. Pythagorean cosmologist, astronomer, mathematician and medical theorist: his medical and physiological doctrines are reported in Anonymus Londinensis.

PLATO OF ATHENS. 428–347 B.C. Influential both for his epistemological views and for the teleological cosmology (which includes a detailed theory of diseases) in the *Timaeus*.

POLYBUS. *Fl.* fourth century B.C. Son-in-law of Hippocrates and probable author of *The Nature of Man* in the Hippocratic collection.

PRAXAGORAS OF COS. *Fl. c.* 300 B.C. Physician and medical theorist, the discoverer of the diagnostic value of the pulse.

RUFUS OF EPHESUS. *Fl. c.* A.D. 100. Medical writer. The principal work of his that is extant is his *On the Naming of the Parts of Man*.

SORANUS OF EPHESUS. Physician who worked about 120 A.D. The most important work that is extant is his treatise in four books on gynaecology. A Life of Hippocrates is also attributed to him.

THEOPHILUS PROTOSPATHARIUS. Seventh-century A.D. medical writer, several of whose works (such as the *On the Construction of the Human Body*, based largely on Galen) are extant.

GENERAL INDEX

SUPPLEMENTARY INDEX TO
MEDICAL TREATISES

This index lists certain morbid conditions which, although not necessarily mentioned by name in the text, are certainly, probably or possibly described therein. It must be realized that an index of this kind, depending on interpretation and sometimes reasonable guesswork, must be liable to a good deal of error; nevertheless it may add some interest for the modern student who may look up such subjects as *mumps* or *diphtheria* or the *inheritance of acquired characteristics* and find something about these subjects in the text. The index should be used in conjunction with the General Index. Where a cross reference is marked with an asterisk (*) the subject will be found in the General Index. Textual references are printed in italic type, page numbers in roman.

The following abbreviations are used:

TM	*Tradition in Medicine*
E(I)	*Epidemics*, book I
E(III)	*Epidemics*, book III
SM	*The Science of Medicine*
AWP	*Airs, Waters, Places*
P	*Prognosis*
A	*Aphorisms*
SD	*The Sacred Disease*

THE STORY OF PENGUIN CLASSICS

Before 1946 ... 'Classics' are mainly the domain of academics and students; readable editions for everyone else are almost unheard of. This all changes when a little-known classicist, E. V. Rieu, presents Penguin founder Allen Lane with the translation of Homer's *Odyssey* that he has been working on in his spare time.

1946 Penguin Classics debuts with *The Odyssey*, which promptly sells three million copies. Suddenly, classics are no longer for the privileged few.

1950s Rieu, now series editor, turns to professional writers for the best modern, readable translations, including Dorothy L. Sayers's *Inferno* and Robert Graves's unexpurgated *Twelve Caesars*.

1960s The Classics are given the distinctive black covers that have remained a constant throughout the life of the series. Rieu retires in 1964, hailing the Penguin Classics list as 'the greatest educative force of the twentieth century.'

1970s A new generation of translators swells the Penguin Classics ranks, introducing readers of English to classics of world literature from more than twenty languages. The list grows to encompass more history, philosophy, science, religion and politics.

1980s The Penguin American Library launches with titles such as *Uncle Tom's Cabin*, and joins forces with Penguin Classics to provide the most comprehensive library of world literature available from any paperback publisher.

1990s The launch of Penguin Audiobooks brings the classics to a listening audience for the first time, and in 1999 the worldwide launch of the Penguin Classics website extends their reach to the global online community.

The 21st Century Penguin Classics are completely redesigned for the first time in nearly twenty years. This world-famous series now consists of more than 1300 titles, making the widest range of the best books ever written available to millions – and constantly redefining what makes a 'classic'.

The Odyssey continues ...

The best books ever written

PENGUIN CLASSICS

SINCE 1946

Find out more at www.penguinclassics.com